D0806204

Inventing Our Selves provides a radical new approach to the analysis of our current regime of the self, and the values of autonomy, identity, individuality, liberty, and choice that animate it. It draws upon the work of Michel Foucault, Gilles Deleuze, and recent feminist scholarship on the body and the self to propose a novel genealogy of subjectivity. It argues that the "psy" disciplines – psychology in particular – have played a key role in "inventing our selves," making visible and practicable certain features of persons, their conducts and their relations with one another, inventing new forms of expertise, transforming authority in a therapeutic direction, and changing the ethical techniques by means of which humans have come to understand and act upon themselves in the name of their truth. This is illustrated through studies of psy disciplines in factories, schools, clinics, the military, public opinion, and therapy. Nikolas Rose argues that the proliferation of psy has been intrinsically linked with transformations in "governmentality," in the rationalities and technologies of political power in contemporary liberal democracies. The aim of this critical history is to diagnose our contemporary condition of the self, to destabilize and denaturalize what seems immutable, to elucidate the burdens imposed, the illusions entailed, the acts of domination and self-mastery that are the counterpart of the capacities and liberties that make up the contemporary individual.

Inventing our selves

Cambridge Studies in the History of Psychology

GENERAL EDITORS

MITCHELL G. ASH AND WILLIAM R. WOODWARD

This new series provides a publishing forum for outstanding scholarly work in the history of psychology. The creation of the series reflects a growing concentration in this area by historians and philosophers of science, intellectual and cultural historians, and psychologists interested in historical and theoretical issues.

The series is open both to manuscripts dealing with the history of psychological theory and research and to work focusing on the varied social, cultural, and institutional contexts and impacts of psychology. Writing about psychological thinking and research of any period will be considered. In addition to innovative treatments of traditional topics in the field, the editors particularly welcome work that breaks new ground by offering historical considerations of issues such as the linkages of academic and applied psychology with other fields, for example, psychiatry, anthropology, sociology, and psychoanalysis; international, intercultural, or gender-specific differences in psychological theory and research; or the history of psychological research practices. The series will include both single-authored monographs and occasional coherently defined, rigorously edited essay collections.

Also in the series

Gestalt psychology in German culture, 1890–1967: Holism and the quest for objectivity
MITCHELL G. ASH

Constructing the subject: Historical origins of psychological research
KURT DANZIGER

Changing the rules: Psychology in the Netherlands, 1900–1985
TRUDY DEHUE

The professionalization of psychology in Nazi Germany
ULFRIED GEUTER

Crowds, psychology, and politics, 1871–1899
JAAP VAN GINNEKEN

Metaphors in the history of psychology
edited by DAVID E. LEARY

Inventing our selves

Psychology, power, and personhood

Nikolas Rose
Goldsmiths College,
University of London

CAMBRIDGE
UNIVERSITY PRESS

PUBLISHED BY THE PRESS SYNDICATE OF THE UNIVERSITY OF CAMBRIDGE
The Pitt Building, Trumpington Street, Cambridge CB2 1RP

CAMBRIDGE UNIVERSITY PRESS
The Edinburgh Building, Cambridge CB2 2RU, United Kingdom
40 West 20th Street, New York, NY 10011-4211, USA
10 Stamford Road, Oakleigh, Melbourne 3166, Australia

First published 1996
First paperback edition 1998

Printed in the United States of America

Library of Congress Cataloging-in-Publication Data is available.

A catalog record for this book is available from the British Library.

ISBN 0 521 43414 9 hardback
ISBN 0 521 64607 3 paperback

Contents

Contents

Acknowledgments

The essays collected in this volume were written over a ten-year period from 1984 to 1994. In the original writing of the essays, and in developing them for publication in this form, I have been assisted by many people, and I would like to thank all those who have commented, criticized, and argued with the approach I have been advocating. I would like to express particular appreciation to three people: Peter Miller, with whom I have collaborated closely for many years on projects adjacent to the work presented here, and whose ideas have greatly enriched my own; Thomas Osborne, who has read, criticized, provoked, and supported my work over the past five years; and Diana Rose, who has, as ever, educated me on psychology, forced me to clarify my arguments, and given me some sense that this work is worth doing.

This volume was largely assembled while I was a Visiting Fellow in the Political Science Program of the Research School for Social Sciences of the Australian National University, and I would like to thank that institution, and my temporary colleagues in Canberra, for providing a most intellectually stimulating and congenial space for writing. In particular I would like to thank Barry Hindess for arranging my visit and for combining challenging discussion with generous hospitality. I would also like to thank my hosts in Melbourne, Deborah Tyler and David McCallum, and those who made me welcome in Brisbane, especially Jeffrey Minson, Denise Meredyth, and others at the School of Humanities at Griffith University for many stimulating conversations and happy hours. The final manuscript could not have been put together on schedule without the help of Diana Lee Woolf of Goldsmiths College.

Earlier versions of the material presented in Chapters 1 and 2 were published as "Identity, genealogy, history," in S. Hall and P. du Gay, eds., *Questions of Cultural Identity* (London: Sage, 1995), and "Power and subjectivity: Critical history and psychology," in K. Gergen and C. Grauman, eds., *His-*

torical Dimensions of Psychological Discourse (Cambridge: Cambridge University Press, 1995). A different version of the argument in Chapter 1 was previously published in "Authority and the genealogy of subjectivity," in S. Lash, P. Heelas, and P. Morris, eds., *De-traditionalization: Authority and Self in an Age of Cultural Uncertainty* (Oxford: Blackwell, 1995). Chapter 2 also draws upon some parts of "Calculable minds and manageable individuals," *History of the Human Sciences,* 1 (1988): 179–200.

An earlier version of Chapter 3 was published as "Psychology as a 'social' science," in I. Parker and J. Shotter, eds., *Deconstructing Social Psychology* (London: Routledge, 1989), pp. 103–16.

The original version of Chapter 4 was presented at the 9th Cheiron-Europe Conference, Weimar, 4–8 September 1990. I have benefited from the comments made by those at the conference, and from the advice of an anonymous reviewer for *Science in Context,* where an earlier version was published as "Engineering the human soul: Analyzing psychological expertise," *Science in Context,* 5, 2 (1992): 351–69.

The material in Chapter 5 was first presented at the International Conference for the History of the Human Sciences, University of Durham, September 1986. A rather different version of some of the same material was presented at a Symposium of the Group for the History of Psychiatry, Psychology and Allied Sciences at the University of Cambridge, September 1986. I would like to acknowledge the indebtedness of this paper to the work of Bruno Latour and Michael Lynch. Thanks also to Roger Smith and reviewers for the *History of the Human Sciences* for advice in relation to the version of the paper that was published there as "Calculable minds and manageable individuals," *History of the Human Sciences,* 1 (1988): 179–200.

Chapter 6 is a revised and much extended version of a paper delivered at the 8th Cheiron-Europe Conference, Göteborg, Sweden, 30 August–3 September 1989. It also draws on arguments developed in *Governing the Soul: The Shaping of the Private Self* (London: Routledge, 1990), and formulated in the course of my work with Peter Miller on the Tavistock Clinic and Tavistock Institute of Human Relations. I would like to thank Diana Rose for her advice on social psychology during the preparation of this paper.

The argument in Chapter 7 was initially presented at a conference on the "The Values of the Enterprise Culture" held at Lancaster University in 1989 and published as "Governing the enterprising self," in Paul Heelas and Paul Morris, eds., *The Values of the Enterprise Culture: The Moral Debate* (London: Routledge, 1992), pp. 141–64. Paul Heelas and Paul Morris gave helpful comments at an early stage.

Chapter 8 draws upon a variety of my published papers, but is presented in this version for the first time. Thanks to Mariana Valverde for her stimulating comments, which helped me refine the arguments in this chapter, and to Thomas Osborne, whose advice on a first draft of this chapter saved me from making even more errors of judgment than are undoubtedly contained in what I have written here.

Introduction

If there is one value that seems beyond reproach, in our current confused ethical climate, it is that of the self and the terms that cluster around it – autonomy, identity, individuality, liberty, choice, fulfillment. It is in terms of our autonomous selves that we understand our passions and desires, shape our life-styles, choose our partners, marriage, even parenthood. It is in the name of the kinds of persons that we really are that we consume commodities, act out our tastes, fashion our bodies, display our distinctiveness. Our politics loudly proclaims its commitment to respect for the rights and powers of the citizen as an individual. Our ethical dilemmas are debated in similar terms, whether they concern the extension of legal protections to same-sex couples, disputes over abortion, or worries about the new reproductive technologies. In less parochial domains, notions of autonomy and identity act as ideals or criteria of judgment in conflicts over national identities, in struggles over the rights of minorities, and in a whole variety of national and international disputes. This ethic of the free, autonomous self seems to trace out something quite fundamental in the ways in which modern men and women have come to understand, experience, and evaluate themselves, their actions, and their lives.

In writing the essays that are collected in this volume, I wanted to make a contribution, both conceptual and empirical, to the genealogy of this current regime of the self. I hope that they will make a modest contribution to our understanding of the conditions under which our present ways of thinking about and acting upon human beings have taken shape; that they will help us chart their characteristic modes of operation; that they will assist us to draw up some kind of evaluation of the capacities they attribute to us and the demands they make of us. My aim, in other words, is to begin to question some of our contemporary certainties about the kinds of people we take ourselves to be, to help develop ways in which we might begin to think ourselves otherwise.

These studies try to problematize our contemporary regime of the self by examining some of the processes through which this regulative ideal of the self has been invented. The invention in question is a historical rather than an individual phenomenon. Hence this work is underpinned by the belief that historical investigation can open up our contemporary regime of the self to critical thought, that is to say, to a kind of thought that can work on the limits of what is thinkable, extend those limits, and hence enhance the contestability of what we take to be natural and inevitable about our current ways of relating to ourselves. The psychosciences and disciplines – psychology, psychiatry, and their cognates – form the focus of these studies. Collectively I refer to the ways of thinking and acting brought into existence by these disciplines since the last half of the nineteenth century as 'psy', not because they form a monolithic or coherent bloc – quite the reverse – but because they have brought into existence a variety of new ways in which human beings have come to understand themselves and do things to themselves. I argue in these essays that psy has played a key role in constituting our current regime of the self as well as itself having been 'disciplinized' as part of the emergence of this regime. However, I do not claim to provide even the sketch for a history of psychology. Rather, I am concerned with the vocabularies, explanations, techniques of psy only to the extent that they bear upon this question of the invention of a certain way of understanding and relating to ourselves and others, to the making of human being intelligible and practicable under a certain description. I want to examine the ways in which the contemporary apparatus for 'being human' has been put together: the technologies and techniques that hold personhood – identity, selfhood, autonomy, and individuality – in place. I term this work 'critical history': its aim is to explore the conditions under which these horizons of our experience have taken shape, to diagnose our contemporary condition of the self, to destabilize and denaturalize that regime of the self which today seems inescapable, to elucidate the burdens imposed, the illusions entailed, the acts of domination and self-mastery that are the counterpart of the capacities and liberties that make up the contemporary individual.

Perhaps it will already be objected that I have set out my question in a misleading fashion, in referring, so hastily, to an experience of oneself in terms such as 'we' and 'our'. Who is this 'we', who comprises this 'our'? Indeed, one of the premises of these essays is that the regime of the self that is prevalent in contemporary Western Europe and North America is unusual both historically and geographically – that its very existence needs to be treated as a problem to be explained. And further, a central argument of these essays is that this regime of the self is indeed more heterogeneous than is often allowed, localized in distinct practices with particular presuppositions about the subjects that inhabit them, varying in its specifications of personhood along a whole number of axes and in different problem spaces – operating differently, for example, in relation to the female murderer, the naughty schoolboy, the young black urban dweller, the depressed housewife of the wealthy classes, the disaffected worker, the redundant middle manager, the entrepreneurial business woman, and so forth. Nonetheless, what justifies me in speaking of a regime of the self, at least within a limited set of temporal and geographical coordinates, is less an assertion of uniformity than a hypothesis that there is a common normativity – a kind of family resemblance in the regulative ideals concerning persons that are at work in all these diverse practices that act upon human beings, young and old, rich and poor, men and women, black and white, prisoner, mad person, patient, boss and worker: ideals concerning our existence as individuals inhabited by an inner psychology that animates and explains our conduct and strives for self-realization, self-esteem, and self-fulfillment in everyday life. The essays that follow should establish the strengths and limits of this hypothesis, and also go some way to trace out the diverse and contingent places, practices, and problems out of which this norm of the quotidian yet sovereign self of choice, autonomy, and freedom has been invented.

To speak of the invention of the self is not to suggest that we are, in some way, the victims of a collective fiction or delusion. That which is invented is not an illusion; it constitutes our truth. To suggest that our relation to ourselves is historical and not ontological is not to suggest that an essential and transhistorical subjectivity lies hidden and disguised beneath the surface of our contemporary experience, as a potential waiting to be realized by means of critique. Nonetheless, these studies do arise out of an unease about the values accorded to the self and its identity in our contemporary form of life, a sense that while our culture of the self accords humans all sorts of capacities and endows all sorts of rights and privileges, it also divides, imposes burdens, and thrives upon the anxieties and disappointments generated by its own promises. I am all too aware that while these essays begin from such an unease, they fall far short of drawing up a balance sheet that would enable us to counterpose the 'costs' of our contemporary experience of our selves against its 'benefits'. I nonetheless hope that, in rendering the historical contingency of our contemporary relations to ourselves more visible, they may help open these up for interrogation and transformation.

The challenged self

The essays have been put together in a time and place in which a series of profound challenges have been directed toward an image of the self that appears, for so long, to have formed the horizon of 'our' thought. The self: coherent, bounded, individualized, intentional, the locus of thought, action, and belief, the origin of its own actions, the beneficiary of a unique biography. As such selves we possessed an identity, which constituted our deepest, most profound reality, which was the repository of our familial heritage and our particular experience as individuals, which animated our thoughts, attitudes, beliefs, and values. As selves, we were characterized by a profound inwardness: conduct, belief, value, and speech were to be interrogated and rendered explicable in terms of an understanding of an inner space that gave them form, within which they were, literally, embodied within us as corporeal beings. This internal universe of the self, this profound 'psychology', lay at the core of those ways of conducting ourselves that are considered normal and provided the norm for thinking and judging the abnormal – whether in the realm of gender, sexuality, vice, illegality, or insanity. And our lives were meaningful, to the extent that we could discover our self, be our self, express our self, love our self, and be loved for the self we really were.

In fact, as I have already hinted, these essays will question whether, or perhaps where, this regulative ideal of the self actually functioned in such a self-evident manner. They will suggest that the images of the person or the subject at work in various practices have historically been more disparate than is implied by such an argument – that diverse conceptions of personhood were deployed in Christian spiritual practices, in the doctor's consulting room, in the hospital operating theater, in erotic relations, in market exchanges, in scholarly activities, in domestic life, in the military. This ideal of the unified, coherent, self-centered subject was, perhaps, most often found in projects that bemoaned the loss of self in modern life, that sought to recover a self, that urged people to respect the self, that urged us each to assert our self and take responsibility for our self – projects whose very existence suggests that selfhood is more an aim or a norm than a natural given. The universal self was reciprocally found in projects to articulate a knowledge of the person, a knowledge structured by the presupposition that an account of the human being had to be, in principle at least, without limits, at least insofar as the humans possessed certain universal characteristics, moral, physiological, psychological, or biological processes that were then worked upon in regular and predictable ways to produce particular and unique individuals. If our current regime of the self has a certain 'systematicity', it is, perhaps, a relatively recent phenomenon, a resultant of all these diverse projects that have sought to know and govern humans *as if they were* selves of certain sorts.

In any event, it is certainly the case that, today, this image of the self has come under question both practically and conceptually. A whole variety of practices bearing upon the mundane difficulties of living a life have placed in question the unity, naturalness, and coherence of the self. The new genetic technology disturbs the naturalness of the self and its boundaries in relation to what is termed, tellingly, its 'reproduction' – donating sperms, transplanting eggs, freezing and implanting embryos, and much more (cf. Strathern, 1992). Abortion and life support machines, together with the contentious debates around them, destabilize the points at which the human enters existence and fades from it. Organ transplants, kidney dialysis, fetal tissue brain implants, heart pacemakers, artificial hearts all problematize the uniqueness of the embodiment of the self, not only establishing 'unnatural' links between different selves via the movement of tissues, but also making all too clear the fact that humans are intrinsically technologically fabricated and 'machinated' – bound into machines in what we term normality as much as in pathology. No wonder that one image of human being has so rapidly disseminated itself: the cyborg (Haraway, 1991).

This image of the human as a cybernetic organism, a nonunified hybrid assembled of body parts and mechanical artifacts, myths, dreams, and fragments of knowledge, is just one dimension of a range of conceptual challenges to the primacy, unity, and givenness of the self. At the very least, within social theory, the idea of the self is historicized and culturally relativized. More radically, it is fractured by gender, race, class, fragmented, deconstructed, revealed not as our inner truth but as our last illusion, not as our ultimate comfort but as an element in circuits of power that make some of us selves while denying full selfhood to others and thus performing an act of domination on both sides.

These contemporary conceptual challenges to the self are, of course, themselves historical and cultural phenomena. As is well known, nineteenth-century social theorists argued in various ways that the process of modernization, the rise of the West, the uniqueness of its values and its economic, legal, cultural, and moral relations could be understood, in part, in terms of 'individualization'. In developing this theme over the course of the twentieth century, and increasingly in its final decades, historians, sociologists, and anthropologists have developed this argument in a different voice, using the historical and cultural specificity of the idea of the self in order to relativize the values of individualism.

The shock value has now faded from assertions such as that by Clifford Geertz that "[t]he Western conception of the person as a bounded, unique, more or less integrated motivational and cognitive universe, a dynamic center of awareness, emotion, judgment and action, organized into a distinctive whole and set contrastively against other such wholes and against a social and natural background is, however incorrigible it may seem to us, a rather peculiar idea within the context of the world's cultures" (Geertz, 1979, p. 229, quoted in Sampson, 1989, p. 1; cf. Mauss, 1979b). In response, the impassioned anthropologist now seeks to retrieve the self from the welter of its social and cultural determinations, and from the relativism that this implies (e.g., Cohen, 1994). But despite such endeavors, it has proved impossible convincingly to reuniversalize and renaturalize this image of the person as a stable, self-conscious, self-identical center of agency.

The peculiarities of our regime of the self have also been diagnosed by philosophers. Historians of philosophy, most notably Charles Taylor, have argued that our modern notion of what it is to be a human agent, a person, or a self, and the issues of morality with which this notion is inextricably intertwined, is "a function of a historically limited mode of self-interpretation, one which has become dominant in the modern West and which may indeed spread thence to other parts of the globe, but which has a beginning in time and space and may have an end" (Taylor, 1989, p. 111). Taylor traces this history through an interpretation of philosophical and literary texts from Plato to the present, seeking to address the "interpretive" question of why people at different historical moments found different versions of the self and identity convincing, inspiring, or moving: the "idée-forces" that is contained within different ideas of the self (p. 203). And Taylor has suggested that our current 'disenchanted' sense of self, in particular the value that we attach to the self that has the capacity to lead, autonomously, an ordinary life, has multiple "sources" arising out of a "theistic" notion that allocates humans souls a special place in the universe, a "romantic" notion stressing the capacity of selves to create and recreate themselves, and a "naturalistic" notion that regards the self as amenable to scientific reason, explicable in terms of biology, heredity, psychology, socialization, and the like. 'The self', whatever virtues of humanity and universality it may entail, thus appears a much more contingent, heterogeneous, culturally relative notion than it purports to be, dependent on a whole complex of other cultural beliefs, values, and forms of life.

Taylor nonetheless retains a certain affection for the regime of the self as it has taken shape historically, and for the moral values to which it has been attached. In this, he is somewhat unusual. The moral valuations underpinning this affection have been most powerfully disputed by feminist philosophers. In different ways, feminists have argued that the cultural representation of the subject *as* a self is based on a continually repeated, motivated, and gendered act of symbolic violence. Beneath the apparent universality of the self as constructed in political thought and philosophy since the seventeenth century lies, in fact, an image of a male subject whose 'universality' is based on its suppressed other. Thus Moira Gatens argues that while the male subject is "constructed as self-contained and as an owner of his person and his capacities, one who relates to other men as free competitors with whom he shares certain politico-economic rights . . . [t]he female subject is constructed as prone to disorder and passion, as economically and politically dependent on men . . . justified by reference to women's nature. She 'makes no sense by herself' and her subjectivity assumes a lack which males complete" (Gatens, 1991, p. 5; cf. Lloyd, 1984). Since its invention, the apparently sex-neutral subject-with-agency was a model applied to one sex and denied to the other; indeed it was dependent on this opposition for its philosophical foundation and political function.

For many who write as feminists, this politicophilosophical and patriarchal illusion of universal 'disembodied' person is to be redressed by an insistence upon the *embodiment* of the subject. The universalizing of the subject, they suggest, went hand in hand with a denial of its bodily existence in favor of a spurious image of reason as abstract, universal, rational, and associated with the masculine principle. A renewed emphasis on embodiment seems to reveal that, at the very least, the subject is two: male and female bodies give rise to radically different forms of subject; the notion of the corporeality of the human is to be developed "by emphasizing the embodied and therefore sexually differentiated structure of the speaking subject" (Braidotti, 1994a, p. 3). Such a reinsertion of 'the body' into our thinking in subjectivity is often argued to have consequences beyond simply questioning the identity of mind and maleness, body and femaleness. For Elizabeth Grosz, if bodies are diverse "male or female, black, brown, white, large or small . . . not as entities in themselves or simply on a linear continuum with its polar extremes occupied by male and female bodies . . . but as a field, a two-dimensional continuum in which race (and possibly even class, caste or religion) form body specifications . . . a defiant affirmation of a multiplicity, a field of differences, of other kinds of bodies and subjectivities . . . if bodies themselves are always sexually (and racially) distinct, incapable of being incorporated into a singular, universal model, then the very forms that subjectivity takes are not generalizable" (Grosz, 1994, p. 19). If subjectivity is understood as corporeal – embodied in bodies that are diversified, regulated according to social protocols, and divided by lines of inequality – then the universalized, naturalized, and rationalized subject of moral philosophy appears in a new light: as the erroneous and troublesome outcome of a denegation of all that is bodily in Western thought.

Feminist theorists have also been at the forefront of a further assault on the image of the unified, individualized psychological self, this time effected through examining the links between subjectification, sexuality, and psychoanalysis. It was Jacques Lacan who began this psychoanalytic assault on the image of the subject that, he suggested, not only infused most contemporary psychology but also the forms of psychoanalysis that had gained sway in the United States and whose regulatory ideal was the mature ego. For Lacan, far from psychoanalysis operating according to the image of harmony and reintegration usually inferred from Freud's dictum that "where id is, there ego shall be," Freud's discovery of the unconscious, and of the rules of its operation, revealed the self's radical ex-centricity to itself. A radical heteronomy gapes within human beings – this is not the property of a few cases of split personality, or a feature of psychological disturbance, but is the very condition of our being able to relate to ourselves as if we were subjects. At the very heart of our assent to our own identity, Lacan asserted, we are wagged, agitated, activated by an Other: an order that goes beyond us and is the condition of any consciousness whatever (Lacan, 1977). Psychoanalysis, in the invention of the notion of the unconscious, is thus taken to have dealt a fundamental blow to the vision of the subject propounded by classical philosophy and taken for granted in everyday existence, by establishing the 'excess' of the subject over its representations of itself. In so doing, it appears, it has made it necessary for us to theorize those psychocultural mechanisms through which the subject comes to take itself as a self.

Again, it is contemporary feminist thought that has pursued these investigations most intensively. With notable exceptions, feminists have insisted that sexual difference is constitutive of subjectivity itself: the identifications that form us *as if* we were subjects are, first of all, articulated in relation to gender (cf. Irigaray, 1985). Thus Judith Butler argues that "the subject, the speaking 'I'" does not precede its construction as gendered, but "is formed by virtue of having gone through such a process of assuming a sex" and that this is a process constitutively bound to exclusion of certain "abject beings" who are not permitted to enjoy the status of a subject in virtue of not according to the forms in which such a sex is prescribed: the existence of such abjected persons, "under the sign of the 'unliveable' is required to circumscribe the domain of the subject" (Butler, 1993, p. 3). Subjectivity, for Butler, is not the origin of action, but the consequence of particular, and inevitably gendered, routines of performativity and modes of citation. The subject and 'its' attributes now appear as an effect of a range of processes that give rise to the human being assuming or taking up a certain position of subject – a position that is not universal but always particular. Subjectification occurs, then, but not in the form in which it thinks itself: subjectivity is no longer unitary, or conceived on the model of the male, but fractured through sexual and racial identifications and regulated by social norms. Yet paradoxically, to account for these practices of subjectification, and to disrupt the ways in which they have traditionally been understood, such arguments themselves seem inescapably drawn to a particular 'theory of the subject' – psychoanalysis – to account for the 'inscription' of the effects of subjectivity in the human animal.

If arguments in anthropology, history, philosophy, feminism, and psychoanalysis have thrown the self into question, they have linked up with arguments developing in the very heartland of the self: the discipline of psychology. For here, too, the self is challenged. For some, the self is to be unsettled by revealing it to be 'socially constructed', 'its attributes' from gender to childhood reconceptualized as multiple and mobile effects of attributions made within historically situated interchanges among people. Thus we are invited "to consider the social origins of the taken-for-granted mind assumptions such as the bifurcation between reason and emotion, the existence of memories, and the symbol system believed to underlie language. [Our attention is directed] to the social, moral, political and economic institutions that sustain and are sustained by current assumptions about human activity" (Gergen, 1985c, p. 5). In these constructivist arguments within psychology, attributions of selfhood and its predicates are most frequently understood in Wittgensteinian terms, as features of language games arising within and making possible certain forms of life: it is in and through language, and only in and through language, that we ascribe to ourselves bodily feelings, intentions, emotions, and all the other psychological attributes that have, for so long, appeared to fill out a natural and given interior volume of the self. "Considered from this point of view, to be a self is not to be a certain kind of being but to be in possession of a certain kind of theory" (Harré, 1985, p. 262; cf. Harré, 1983, 1989).

Either for epistemological reasons (we can never know the inner domain of the person – all we have is language) or for ontological reasons (the entities constructed by psychology do not correspond to the real being of the human), an analysis of a psychological interior is to be replaced by an analysis of the exterior realm of language that *attributes* mental states – beliefs, attitudes, personalities, and the like – to individuals (see the essays collected in Gergen and Davis, 1985, and Shotter and Gergen, 1989). When what was once attributed to a unified psychological domain is now dispersed among culturally diverse linguistic practices, beliefs, and conventions: the unified self is revealed as a construction. Once again the self is challenged and fragmented: heterogeneity is not a temporary condition but the inescapable outcome of the discursive processes through which 'the self' is 'socially constructed'. And, from the perspective of so many of these critical psychological investigations, psychology itself becomes not only a major contributor to contemporary understandings of the person, through the vocabularies and narratives it supplies, but also a discipline whose very existence is to be regarded with suspicion. Why, if human beings are as heterogeneous and situationally produced as they now appear to be, did a discipline arise that promulgated such unified, fixed, interiorized, and individualized conceptions of selves, males and females, races, ages. Whose interests did such an intellectual project serve?

Of course, these contemporary challenges to the coherence of the self, which I have described in barest outline, occupy one dimension of that composite cultural and intellectual movement sometimes termed postmodernism. This has made fashionable the argument that the self, like society and culture, has been transformed in current conditions: subjectivity is now fragmented, multiple, contradictory, and the human condition entails each of us trying to make a life for ourselves under the constant gaze of our own suspicious reflexivity, tormented by uncertainty and doubt. I think we would be well to treat these breathless pronouncements of the uniqueness of our age and our special position in history – we are at the end of something, at the start of something – with a certain reserve. In the essays that follow, drawing upon many of the ideas that I have mentioned, I suggest some pathways for a more sober critical assessment of the birth and functioning of our contemporary regime of the self. The multiplicity of regimes of subjectification is not, I suggest, a novel feature of our own age. The repetition of the parameters of difference – gender, race, class, age sexuality, and the like – may perform a useful polemical function, but at most such parameters gesture to the starting points for an analysis of modes of subjectification, not to its conclusions: these categories, too, have a history and a location within particular practices of the person. 'The body' provides no sure basis for an analytic of subjectification, precisely because corporealities are diverse, nonunified, and operate in relation to particular regimes of knowledge: the configurations of the human body inscribed in the anatomical atlas did not always define a way of delimiting the order of vital processes, or of visualizing and acting upon human being. The binary division of gender imposes a fallacious unification on a diversity of ways in which we are 'sexed' – as men, women, boys, girls, manly, feminine, blokes, perverts, homosexuals, gay, lesbian, seducers, mistresses, lovers, ladies, matrons, spinsters. No theory of the psyche can provide the basis for a genealogy of subjectification, precisely because the emergence of such theories has been central to the very regime of the self whose birth must be the object of our inquiries. The notion of interests as explaining the positions espoused in intellectual and practical disputes is inadequate, because what is involved is the *creation* of 'interests', the forging of novel relations between knowledge and politics, and the association and mobilization of forces around them. And, while there is much of value in the attention directed by critical psychology to the conditions of the birth and functioning of the discipline, the focus on language and narrative, on the subjectification as a matter of the stories we tell ourselves about ourselves, is, at best, partial, at worst misguided. Subjectification is not to be understood by locating it in a universe of meaning or an interactional context of narratives, but in a complex of apparatuses, practices, machinations, and assemblages within which human being has been fabricated, and which presuppose and enjoin particular relations with ourselves. Such, at least, will be the argument developed, in different ways, in this book.

Subjectification: Government and psy

These studies arise at the intersection of two concerns that appear to me to be intrinsically related. The first of these is a concern with the history of psychology, or rather, of all those disciplines which, since about the middle of the nineteenth century, have designated themselves with the prefix psy – psychology, psychiatry, psychotherapy, psychoanalysis. This may seem perverse and limiting, for the psy disciplines are, after all, only a small element in contemporary culture, little understood by most people. Indeed, in popular culture, where psy is not parodied, it is often represented – or 'misrepresented' – in a way that makes professional and academic practitioners of the psychological specialisms throw up their hands in exasperation. However, I want to suggest that psychology, in the sense in which I will use the term here, has played a rather fundamental part in 'making up' the kinds of persons that we take ourselves to be. Psychology, in this sense, is not a body of abstracted theories and explanations, but an 'intellectual technology', a way of making visible and intelligible certain features of persons, their conducts, and their relations with one another. Further, psychology is an activity that is never purely academic; it is an enterprise grounded in an intrinsic relation between its place in the academy and its place as 'expertise' (Danziger, 1990). By expertise is meant the capacity of psychology to provide a corps of trained and credentialed persons claiming special competence in the administration of persons and interpersonal relations, and a body of techniques and procedures claiming to make possible the rational and human management of human resources in industry, the military, and social life more generally.

In these essays I argue that the growth of the intellectual and practical technologies of psychology in Europe and North America over the period since the late nineteenth century is intrinsically linked with transformations in the exercise of political power in contemporary liberal democracies. And I also suggest that the growth of psy has been connected, in an important way, with transformations in forms of personhood – our conceptions of what persons are and how we should understand and act toward them, and our notions of what each of us is in ourselves, and how we can become what we want to be. In posing the matter in this way, my investigations take their inspiration from the writings of Michel Foucault. They are attempts to explore "the games of truth and error through which being is historically constituted as experience; that is as something that can and must be thought" (Foucault, 1985, pp. 6–7). By experience here, Foucault does not refer to something primordial that precedes thought, but to "the correlation between fields of knowledge, types of normativity, and forms of subjectivity in a particular culture" (p. 3), and it is in something like this sense that I use the term in this book. I explore aspects of the regimes of knowledge through which human beings have come to recognize themselves as certain kinds of creature, the strategies of regulation and tactics of action to which these regimes of knowledge have been connected, and the correlative relations that human beings have established with themselves, in taking themselves as subjects. In so doing, I hope to contribute to the type of work that Foucault described as an analysis of "the *problematizations* through which being offers itself to be, necessarily, thought – and the *practices* on the basis of which these problematizations are formed" (p. 11).

From this perspective, the history of the psy disciplines is much more than a history of a particular and often somewhat dubious group of sciences – it is part of the history of the ways in which human beings have regulated others and have regulated themselves in the light of certain games of truth. But, on the other hand, this regulatory role of psy is linked, I suggest, to questions of the organization and reorganization of political power that have been quite central to shaping our contemporary experience. The history of psy, that is to say, is intrinsically linked to the history of government. By government I do not just mean politics, although I will argue in the studies that follow that psy knowledge, techniques, explanations, and experts have often entered directly into the concerns, deliberations, and strategies of politicians and others directly linked to the political apparatus of the state, civil and public services, welfare, and so forth, as well as those politically concerned with the organization of military and economic affairs. But I use the term government in the much broader sense given to it by Foucault – government here is a way of conceptualizing all those more or less rationalized programs, strategies, and tactics for 'the conduct of conduct', for acting upon the actions of others in order to achieve certain ends (Foucault, 1991; see Rose 1990, Miller and Rose, 1990, Rose and Miller, 1992). In this sense one might speak of the government of a ship, of a family, of a prison or factory, of a colony, and of a nation, as well as of the government of oneself.

The perspective of government draws our attention to all those multitudinous programs, proposals, and policies that have attempted to shape the conduct of individuals – not just to control, subdue, discipline, normalize, or reform them, but also to make them more intelligent, wise, happy, virtuous, healthy, productive, docile, enterprising, fulfilled, self-esteeming, empowered, or whatever. It helps us free ourselves from the profoundly misleading view that we should understand the practices of normativity that have shaped our present in terms of the political apparatus of the state. The state and the political are relocated as shifting zones for the coordination, codification, and legitimation of some of the complex and diverse array of practices for the government of conduct that exist at a particular time and place. In practices for the government of conduct, from those of social insurance to those of industrial management, those of social and individual hygiene to those of family-focused social work, authorities abound whose powers are based on their professional training and their possession of esoteric ways of understanding and acting upon conduct grounded in codes of knowledge and claims to special wisdom. This perspective draws our attention to the role of knowledge in the contemporary conduct of conduct – where any legitimate attempt to act upon conduct must embody some way of understanding, classifying, calculating, and hence be articulated in terms of some more or less explicit system of thought and judgment. And it also emphasizes the fact that, in the history of power relations in liberal and democratic regimes, the government of others has always been linked to a certain way in which 'free' individuals are enjoined to govern themselves as subjects simultaneously of liberty and of responsibility – prudence, sobriety, steadfastness, adjustment, self-fulfillment, and the like.

It is in this sense that the history of psychology in liberal societies joins up with the history of liberal government. In the studies that follow, I shall examine the ways in which these twin paths are intertwined, in which the historical development, transformation, and proliferation of psy has been bound up with the transformations in rationalities for government, and in the technologies invented to govern conduct. In one sense, psychological experts have a role in this history that is not unique – one could also trace the part played by a whole variety of other 'specialists' in the shaping of the ways that human beings have come to experience themselves: lawyers, economists, accountants, sociologists, anthropologists, political scientists. Indeed all the experts of the human sciences including, for instance, those who have studied animal behavior, physiology, demography, the epidemiology of disease, and human geography have undoubtedly played a part in establishing these practices of recognition. However, in another sense, or so I shall claim, psy experts have achieved a certain privileged position over the past century – for it is psy that claims to understand the inner determinants of human conduct, and psy that thus asserts its ability to provide the appropriate underpinning, in knowledge, judgment, and technique, for the powers of experts of conduct wherever they are to be exercised.

Thus these are studies in the ways in which persons have been invented – 'made up', as Ian Hacking has put it, at the multitude of points of intersection between practices for the government of others and techniques for the government of oneself (cf. Hacking, 1986). They advance the thesis that the growth of the intellectual and practical technologies of psy is intrinsically linked with transformations in the practices for 'the conduct of conduct' that have been assembled in contemporary liberal democracies. Of course, the history of the psy sciences cannot be reduced to their capacities to make human being governable; the complex and heterogeneous process of formation and reformation of disciplines and systems of thought do not inevitably have regulatory aspirations either as their conscious aim or as their covert determinant. But, I suggest, this history is not intelligible without taking account of the complex relations between problems of governability and the invention, stabilization, and institutionalization of psy knowledges. In this process, new configurations have been given, not only to the nature of authority and to the relations that authorities have with their subjects, but also to our relations with ourselves. In particular, I suggest that the novel forms of government being invented in so many 'postwelfare' nations at the close of the twentieth century have come to depend, perhaps as never before, upon instrumentalizing the capacities and properties of 'the subjects of government', and therefore cannot be understood without addressing these new ways of understanding and acting upon ourselves and others as selves 'free to choose'.

To argue that psy has played a constitutive role in the practices of subjectification that are vital to the governability of liberal democracy is not, of course, to suggest that psychologists, psychiatrists, and psy technologies played no role in authoritarian strategies of government. The nineteenth-century reformatory institutions that provided key conditions for the birth of psy were vital elements in governmental strategies that saw the obligatory inculcation of discipline into each citizen as a necessary subjective condition for the establishment of liberty (Foucault, 1977; Rose, 1993). In the late nineteenth and early twentieth century, in Britain and the United States, the expertise of psy was intrinsically bound to eugenic strategies in which the liberty of the majority was to be safeguarded by coercively constraining the reproductive capacities, freedom of movement, and even the life of all those who would threaten the well-being of the race (I discuss eugenics at length in Rose, 1985). More pertinently, perhaps, in Nazi Germany, in the Soviet Union, and in the Communist states of Eastern Europe, psychological and psychiatric expertise was certainly called upon to play its part in the regulation of individuals and populations.

A number of excellent studies of the history of psychology in Nazi Germany have illuminated these relationships (Ash, 1995; Geuter, 1992; Cocks, 1985). Key psychological theories of personality, for example, were revised to conform to race theory. German military psychology flourished up until 1941, principally deploying a form of 'characterology' to assess the intelligence, character, strength of will, and ability to command of potential officers; for reasons that remain unclear, military psychology was suddenly disbanded in 1942. The powers of psy in the school, the courtroom, the factory, and other such institutional domains had, in fact, already been weakened before the rise of the Nazis. The spread of the technologies of quantification and measurement of human capacities, so significant in Britain and the United States, was limited by the success of those who advocated methods that would address feeling, the will, and experience – methods such as handwriting analysis, which would not treat the individual as merely something measurable and graspable in numbers, but would diagnose the inner powers and structural principles of the human soul. After 1941, the bureaucracy of mass warfare and mass extermination proceeded without much use for the truth claims of the psychosciences or those who professed them (Geuter, 1992). Psychiatry in the Nazi period was drafted into the struggle against biological enemies of the race: a project in which all those deemed to be suffering from severe incurable or congenital mental illness were to be singled out, excluded, sterilized, and finally exterminated. Psychotherapy, rather surprisingly, could be accorded a role under the Nazis in alleviating the mental distress of members of the German *Volksgemeinschaft,* but it did not become a widely deployed technology for the regulation of conduct or subjectivity (Cocks, 1985). Thus the relations of Nazi government and psy were complex and ambivalent. Geuter concludes that, while many psychologists did try to place their discipline in the service of organs of Nazi domination, psychology contributed little to stabilizing that domination – its role in officer selection for the Wehrmacht was neither necessary nor particularly significant, it was not systematically involved in the development of official propaganda, and psychologists are not known to have been used by the Nazis or the SS in persecution, torture, or murder (Geuter, 1992). Although the institutional involvement of psy during the Nazi period did provide some of the key conditions for its later professionalization, Nazi rule did not appear to require a neutral, rational, technical expertise of subjectivity.

The psy disciplines also played a role in communist states. The role of psychiatry in the confinement of political dissidents in the Soviet Union from the late 1930s onward is well known (Bloch and Redaway, 1977; United States Congress, 1973). This was, however, only one element within a strategy set in place following the Bolshevik Revolution that sought to deploy psychiatry across the territory of society in the attempt to prevent mental illness, utilizing reeducative tactics to return malfunctioning citizens to social normality and industrial productivity. Other studies of soviet psychology have similarly suggested that the psy disciplines, at certain historical moments, were significant in the regulation of the conduct of the new Soviet citizen – whether as a schoolchild, as a laborer, or as a disturbed member of society (Wortis, 1950; Bauer, 1952; Rollins, 1972; Kozulin, 1984; Joravsky, 1989; for the internal history of soviet psychology, see Cole and Maltzman, 1969). The new communist citizen was accorded a particular subjectivity, and this subjectivity was connected up in key ways to the development and deployment of psy. For example, psychotechnics was utilized in industry in the 1920s and 1930s, and the International Psychotechnics Congress of 1931 was held in the Soviet Union, although the distinction was stressed between bourgeois psychotechnics, based on the premise of the immutability of abilities and designed to perpetuate the class order and the oppression of minorities, and soviet psychotechnics which placed emphasis on techniques of training which could mold and reshape the laborer to meet the demands for skilled workers in the expanding economy (Bauer, 1952, p. 107). Further, psychological expertise was widely deployed in the service of progressivist pedagogy in shaping the practices of education and in the assessment of pupils. In the same period, the study of attitudes flourished briefly, and psychological tests were used for officer selection in the Red Army.

In the second half of the 1930s, almost all of these psy strategies for the government of the human factor were halted. A decree of 1936 from the Central Committee of the Communist Party abolished pedagogy: the schoolroom was now to be governed according to an older regime of military discipline, habit formation, and hierarchical instructions (Kozulin, 1984, pp. 121–36). Psychological testing in the Soviet Union was banned: tests were branded as reactionary, bourgeois instruments designed to perpetuate the class structure, and contrary to the principle of reeducation, which was central to the practice of Socialist reconstruction (Bauer, 1952, pp. 116–27). Similarly, it was declared that attitude questionnaires that concerned the subject's political views or "probed into the deeper and intimate side of life must be *categorically banned*" (Bauer, 1952, p. 111). Although there was undoubtedly a rebirth of psychology after World War II, the governmental role of psy expertise in postwar communist nations remains to be analyzed. From the few detailed studies of the local party apparatus that are available, there is little evidence that the experts of psy were of much importance in the 'pastoral' relations of the Communist Party bureaucracies through which everyday life was regulated in the former communist states of Eastern Europe in the period preceding their collapse (Horvath and Szakolczai, 1992).

If I do not discuss these relations between nonliberal governmentality and psy in these essays, this is not because I consider them insignificant. I certainly do not want to argue that psy knowledges and techniques have any necessary political allegiance or destiny, far less a liberal one. The argument that follows has a more limited scope, but one that may be most useful for understanding the political and ethical dilemmas that arise today as the slogans of freedom and autonomy spread across Central and Eastern Europe, China, and the other regions that are being opened up to the penetration of free-market economics, anticollectivist cultural politics, and the technologies of consumption. What I wish to show in these studies is that, over a particular and limited historical period and geographical dispersion, the languages, techniques, forms of expertise, and modes of subjectification constitutive of modern liberal democracies – indeed, of the very meaning of life itself – have been made possible by, and shaped by, the modes of thinking and acting that I term psy. Most crucially, I suggest, psy has infused the shape and character of what we take to be liberty, autonomy, and choice in our politics and our ethics; in the process, freedom has assumed an inescapably subjective form.

What is at stake in these analyses at their most general, therefore, is nothing less than freedom itself: freedom as it has been articulated into norms and principles for organizing our experience of our world and of ourselves; freedom as it is realized in certain ways of exercising power over others; freedom as it has been articulated into certain rationales for practicing in relation to ourselves (cf. Rose, 1993, for what follows). How have we come to define and act toward ourselves in terms of a certain notion of freedom? How has freedom provided the rationale for all manner of coercive interventions into the lives of those seen as unfree or threats to freedom: the poor, the homeless, the mad, the risky, or those at risk? What are the relations between rationalities and techniques of government that have sought to justify themselves in terms of freedom and these practices of the self regulated by norms of freedom? These studies suggest that at least one central feature of the emergence of this contemporary regime of the free individual, and the political rationalities of liberalism to which freedom is so dear, has been the invention of a range of psy technologies for governing individuals *in terms of their freedom.* The importance of liberalism as an ethos of government, rather than as political philosophy, is thus not that it first recognized, defined, or defended freedom as a right of all citizens. Rather, its significance is that for the first time the arts of government were systematically linked to the practice of freedom and hence to the characteristics of human beings as potentially subjects of freedom. From this point on, to quote John Rajchman, individuals "must be willing to do their bit in maintaining the systems that define and delimit them; they must play their parts in a 'game' whose intelligibility and limits they take for granted" (Rajchman, 1991, p. 101). The forms of freedom we inhabit today are intrinsically bound to a regime of subjectification in which subjects are not merely 'free to choose', but *obliged to be free,* to understand and enact their lives in terms of choice under conditions that systematically limit the capacities of so many to shape their own destiny. Human beings must interpret their past, and dream their future, as outcomes of personal choices made or choices still to make yet within a narrow range of possibilities whose restrictions are hard to discern because they form the horizon of what is thinkable. Their choices are, in their turn, seen as realization of the attributes of the choosing self – expressions of personality – and reflect back upon the individual who has made them. The practice of freedom appears only as the possibility of the maximum self-fulfillment of the active and autonomous individual.

In the nineteenth century, psychology invented the normal individual. In the first half of this century it was a discipline of the social person. Today, psychologists elaborate complex emotional, interpersonal, and organizational techniques by which the practices of everyday life can be organized according to the ethic of autonomous selfhood. Correlatively, freedom has come to mean the realization of the potentials of the psychological self in and through activities in the mundane world of everyday life. The significance of psychology, here, is the elaboration of a know-how of this autonomous individual striving for self-realization. Psychology has thus participated in reshaping the practices of those who exercise authority over others – social workers, managers, teachers, nurses – such that they nurture and direct these individual strivings in the most appropriate and productive fashions. It has invented what one might term the therapies of normality or the psychologies of everyday life, the pedagogies of self-fulfillment disseminated through the mass media, which translate the enigmatic desires and dissatisfactions of the individual into precise ways of inspecting oneself, accounting for oneself, and working upon oneself in order to realize one's potential, gain happiness, and exercise one's autonomy. And, it has given birth to a range of psychotherapies that aspire to enabling humans to live as free individuals through subordinating themselves to a form of therapeutic authority: to live as an autonomous individual, you must learn new techniques for understanding and practicing upon yourself. Freedom, that is to say, is enacted only at the price of relying upon experts of the soul. We have been freed from the arbitrary prescriptions of religious and political authorities, thus allowing a range of different answers to the question of how we should live. But we have been bound into relationship with new authorities, which are more profoundly subjectifying because they appear to emanate from our individual desires to fulfill ourselves in our everyday lives, to craft our personalities, to discover who we really are. Through these transformations we have 'invented ourselves' with all the ambiguous costs and benefits that this invention has entailed.

The structure of this book

How should one do the history of the self? In the first chapter of this book, I argue that we should not seek to answer this question by writing a history of the person in which individuality and individualism function as key events in a transition to 'modernity' and in which our present features as the moment of a similar pivotal historical transformation in personhood. Rather I suggest an approach that I term 'the genealogy of subjectification', a genealogy of our modern regime of the self, of our 'relation to ourselves' that takes the interiorized, totalized, and psychologized understanding of what it is to be human as the site of a historical problem. I suggest an approach to the analytics of this relation with ourselves that focuses upon the *practices* within which human beings have been addressed and located. Such an analysis proceeds along a number of linked pathways: problematizations, technologies, authorities, teleologies, and strategies. I argue that regimes of subjectification are heterogeneous and that this heterogeneity is significant in relation to their modes of functioning. The heterogeneity, and the practical embededness of regimes of subjectification, enables us to account for the omnipresence of conflict, agency, and resistance without positing some essential subjectivity or desire. I also make some suggestions as to how a genealogy of subjectification might conceptualize the human material upon which history writes, suggesting some elements of a minimal, weak, or thin conception of human being.

There has recently been something of a revival of historical work on psychology, carried out under a number of different theoretical auspices. In Chapter 2 I argue for a particular approach to the history of psychology that I term 'critical history'. A critical history, I suggest, is one that helps us think about the nature and limits of our present, about the conditions under which that which we take for truth and reality has been established. Critical history disturbs and fragments, it reveals the fragility of that which seems solid, the contingency of that which seems necessary, the mundane and quotidian roots of that which claims lofty nobility and disinterest. It enables us to think *against* the present, in the sense of exploring its horizons and its conditions of possibility. Its aim is not to predetermine judgment, but to make judgment possible. In this chapter, I counterpose my perspective on the relations between the psychological, the social and the subjective to other recent approaches to the history of psychology, society, and the subject, arguing that psychology should not be viewed as an effect or instrument of power, but as a domain that is 'disciplined' in relation to certain practices and problems of government, is dependent for its epistemology on certain institutional forms and regimes of judgment in relation to human conduct, and as that 'know-how' which makes certain 'power effects' possible. Psychology, and all the psy knowledges, has played a significant role in the reorganization of the practices and techniques that have linked authority to subjectivity over the past century, especially in the liberal and democratic polities of Europe, the United States, and Australia.

Radical critics often imply that psychology is committed to 'individualism', and that it is therefore an 'antisocial' science. In Chapter 3, I argue that psychology is a profoundly *social* science and even the most 'individualistic' aspects of psychology must be connected into the social field. As the previous chapter shows, psychology is social first of all in that truth is a constitutively social phenomenon. But psychology is social in a second, correlative sense, in that its birth as a distinct discipline – its vocation and its destiny – is intimately linked to those forms of political rationality and governmental technology that have given birth to that domain of our reality that we call 'social' – social security, social work, social insurance. It is this link between psychology and social government that is explored in this chapter.

If it is the case that diverse practices, techniques, and forms of judgment concerning human subjects have become 'psychologized', how should this be understood? In Chapter 4 I propose the concept of *techne* to think about the characteristic ways in which psychology has entered into a range of 'human technologies' – practices seeking certain outcomes in terms of human conduct such as reform, efficiency, education, cure, or virtue. This chapter examines some of these technologies and works toward a classification of them. I argue that psychological modes of thought and action have come to underpin – and then to transform – a range of diverse practices for dealing with persons and conduct that were previously cognized and legitimated in other ways.

Of course, while psychology was formed as a discipline and a specialism in the nineteenth century, reflections on the human psyche have a much longer history. So what differentiates the psychological sciences that were born in the nineteenth century from those discourses on the human soul that preceded them, and how is this difference linked up with other social and political events? What produced their 'disciplinization': the establishment of university departments, professorships, degree programs, laboratories, journals, training courses, professional associations, specialized employment statuses, and so forth? In Chapter 5 I develop Michel Foucault's hypothesis that all the disciplines bearing the prefix psy or psycho have their origin in what he terms a reversal of the political axis of individualization. I examine the role of the psychological sciences as *techniques for the disciplining of human difference:* individualizing humans through classifying them, calibrating their capacities and conducts, inscribing and recording their attributes and deficiencies, managing and utilizing their individuality and variability.

But, as I have already stressed, psychology has a much wider vocation than simply as a 'science of the individual'. In Chapter 6 I examine the links between social psychology and democracy. The social psychology written in the 1930s, 1940s, and 1950s makes frequent references to democracy. These references to democracy are more than rhetorical flourishes. To rule citizens democratically means ruling them through their freedoms, their choices, and their solidarities rather than despite these. It means turning subjects, their motivations and interrelations, from potential sites of resistance to rule into allies of rule. It means replacing arbitrary authority with that permitting rational justification. Social psychology as a complex of knowledges, professionals, techniques, and forms of judgment is constitutively linked to democracy, as a way of organizing, exercising, and legitimating political power. Advanced liberal democratic polities, I suggest, produce some characteristic problems to which the intellectual and practical technologies that comprise social psychology can promise solutions.

What, then, about our contemporary regimes of selfhood and how are they related to current mutations in the field of government? In Chapter 7 I consider the ways in which the current prominence of concerns with the self is bound up with the emergence of a range of political programs and techniques that seek to govern in new ways, not through 'society' but through the educated and informed choices of 'active' citizens, families, and communities. These modes of government accord new powers and roles to authorities and experts and govern the exercise of expert authority in new ways. They seek to operate in accordance with new ways of understanding ourselves, which subjectify in terms of the self-promoting projects of individuals and their families and which instrumentalize the desires for the maximization of quotidian forms of existence in terms of life-style and quality of life. The ethical valorization of certain features of the person – autonomy, freedom, choice, authenticity, enterprise – needs to be understood in terms of new rationalities of government and new technologies for the conduct of conduct. In this sense a critical history of psychology is linked not only to an analytics of our regime of the self, but also to the ethical questions of the costs and benefits of our current relations with ourselves and to the strategies we might invent for governing others, and governing ourselves, otherwise.

In the final chapter of this book, I return to the question of 'subjectivity' itself and how it might be thought differently, unsettled or destabilized. As I have argued in previous chapters, the discourses that have addressed the human person are more than 'representations' of subjective reality or cultural beliefs. They have constituted a body of critical reflections on the problems of governing persons in accordance with, on the one hand, their nature and truth and, on the other, with the demands of social order, harmony, tranquillity, and well-being. They have established arrays of norms according to which the capacities and conduct of the self have been judged. But they have also constituted changing regimes of signification through which persons can accord meaning to themselves and their lives and are manifested in techniques for shaping and reforming selves. While many have announced the 'death of the subject' and proposed alternative models of subjectivity, none have been so provocative in their proposals as Gilles Deleuze and Félix Guattari. In this chapter, after discussing some accounts of the linguistic or narrative construction of subjectivity, I propose a view of subjectification in terms of an array of 'foldings' of exteriority, themselves assembled and machinated in particular apparatuses. Psy, I suggest, has played a key role in the folds through which we, today, have come to relate to ourselves. And an analysis of these psychological 'foldings' goes some way to help us understand how, as inhabitants of this particular spatiotemporal zone, we have been brought to recognize ourselves as subjects of 'freedom'.

1

How should one do the history of the self?

The human being is not the eternal basis of human history and human culture but a historical and cultural artifact. This is the message of studies from a variety of disciplines, which have pointed in different ways to the specificity of our modern Western conception of the person. In such societies, it is suggested, the person is construed as a self, a naturally unique and discrete entity, the boundaries of the body enclosing, as if by definition, an inner life of the psyche, in which are inscribed the experiences of an individual biography. But modern Western societies are unusual in construing the person as such a natural locus of beliefs and desires, with inherent capacities, as the self-evident origin of actions and decisions, as a stable phenomenon exhibiting consistency across different contexts and times. They are also unusual in grounding and justifying their apparatuses for the regulation of conduct upon such a conception of the person. For example, it is in terms of this notion of the self that much of our criminal legal systems operate, with their notions of responsibility and intent. Our systems of morality are similarly historically unusual in their valorization of authenticity and their emotivism. No less unusual, historically, are our politics, which place so much emphasis on individual rights, individual choices, and individual freedoms. It is in these societies that psychology has been born as a scientific discipline, as a positive knowledge of the individual and a particular way of speaking the truth about humans and acting upon them. Further, or so it would appear, in these societies, human beings have come to understand and relate to themselves as 'psychological' beings, to interrogate and narrate themselves in terms of a psychological 'inner life' that holds the secrets of their identity, which they are to discover and fulfill, which is the standard against which the living of an 'authentic' life is to be judged.

How should one write the history of this contemporary 'regime of the self'? I would like to suggest a particular approach to this issue, an approach which

I term 'the genealogy of subjectification'.[1] This phrasing is awkward but, I think, important. Its importance lies, in part, in indicating what such an undertaking is *not.* On the one hand, it is not an attempt to write the history of changing ideas of the person, as they have figured within philosophy, literature, culture, and so on. Historians and philosophers have long engaged in the writing of such narratives, and no doubt they are significant and instructive (e.g., Taylor, 1989; cf. the different approach of Tully, 1993). My concern, however, is not with 'ideas of persons' but with the practices in which persons are understood and acted upon – in relation to their criminality, their health and sickness, their family relations, their productivity, their military role, and so forth. It is unwise to assume that one can derive, from an account of notions of the human being in cosmology, philosophy, aesthetics, or literature, evidence about the presuppositions that shape the conduct of human beings in such mundane sites and practices (cf. Dean, 1994). While a genealogy of subjectification is concerned with human being as it is thought about, therefore, it is not a history of ideas: its domain of investigation is that of practices and techniques, of thought as it seeks to make itself *technical.*

Equally my approach needs to be distinguished from attempts to write the history of the person as a psychological entity, to see how different ages produce humans with different psychological characteristics, different emotions, beliefs, pathologies. Such a project for a history of the person is certainly imaginable and something like this aspiration shapes a number of recent psychological studies, some of which I discuss here. It also animates a number of recent sociological investigations. But such analyses presuppose a way of thinking that is itself an outcome of history, one that emerges only in the nineteenth century. For it is only at this historical moment, and in a limited and localized geographical space, that human being is understood in terms of individuals who are selves, each equipped with an inner domain, a 'psychology', which is structured by the interaction between a particular biographical experience and certain general laws or processes of the human animal.

A genealogy of subjectification takes this individualized, interiorized, totalized, and psychologized understanding of what it is to be human as the site of a historical problem, not as the basis for a historical narrative. Such a genealogy works toward an account of the ways in which this modern regime of the self emerges, not as the outcome of any gradual process of enlightenment, in which humans, aided by the endeavors of science, come at last to recognize their true nature, but out of a number of contingent and altogether less refined and dignified practices and processes. To write such a genealogy is to seek to unpick the ways in which the self that functions as a regulatory ideal in so many aspects of our contemporary forms of life – not merely in our passional relations with one another, but in our projects of life planning, our ways of managing industrial and other organizations, our systems of consumption, many of our genres of literature and aesthetic production – is a kind of 'irreal' plane of projection,[2] put together somewhat contingently and haphazardly at the intersection of a range of distinct histories – of forms of thought, techniques of regulation, problems of organization, and so forth.

Dimensions of our relation to ourselves

A genealogy of subjectification is a genealogy of what one might term, following Michel Foucault, 'our relation to ourselves' (Foucault, 1986b).[3] Its field of investigation comprises the kinds of attention that humans have directed toward themselves and others in different places, spaces, and times. To put this rather more grandly, one might say that this was a genealogy of 'being's relation to itself' and the technical forms that this has assumed. The human being, that is to say, is that kind of creature whose ontology is historical. And the history of human being, therefore, requires an investigation of the intellectual and practical techniques that have comprised the instruments through which being has historically constituted itself: it is a matter of analyzing "the problematizations though which being offers itself to be, necessarily, thought – and the practices on the basis of which these problematizations are formed" (Foucault, 1985, p. 11; cf. Jambet, 1992). The focus of such a genealogy, therefore, is not 'the history of the person' but the genealogy of *the relations* that human beings have established with themselves – in which they have come to relate to themselves *as selves.* These relations are constructed and historical, but they are not to be understood by locating them in some amorphous domain of culture. On the contrary, they are addressed from the perspective of 'government' (Foucault, 1991; cf. Burchell, Gordon, and Miller, 1991). Our relation with ourselves, that is to say, has assumed the form it has because it has been the object of a whole variety of more or less rationalized schemes, which have sought to shape our ways of understanding and enacting our existence as human beings in the name of certain objectives – manliness, femininity, honor, modesty, propriety, civility, discipline, distinction, efficiency, harmony, fulfillment, virtue, pleasure – the list is as diverse and heterogeneous as it is interminable.

One of the reasons for stressing this point is to distinguish my approach from a number of recent analyses that have, explicitly or implicitly, viewed changing forms of subjectivity or identity as consequences of wider social and cultural transformations – modernity, late modernity, the risk society (Bauman, 1991; Beck, 1992; Giddens, 1991; Lash and Friedman, 1992). Of course, this work continues a long tradition of narratives, stretching back at least to Jacob Burckhardt: histories of the rise of the individual as a consequence of a general social transformation from tradition to modernity, feudalism to capitalism, Gemeinschaft to Gesellschaft, mechanical to organic solidarity, and so forth (Burckhardt, [1860] 1990). These kinds of analyses regard changes in the ways in which humans beings understand and act upon themselves as the outcome of 'more fundamental' historical events located elsewhere – in production regimes, in technological change, in alterations in demography or family forms, in 'culture'. No doubt events in each of these domains have significance in relation to the problem of subjectification. But however significant they may be, it is important to insist that such changes do not transform ways of being human by virtue of some 'experience' that they produce. Changing relations of subjectification, I want to argue, cannot be established by derivation or interpretation of other cultural or social forms. To assume explicitly or implicitly that they can is to presume the *continuity* of human beings as the subjects of history, essentially equipped with the capacity for endowing meaning (cf. Dean 1994). But the ways in which humans 'give meaning to experience' have their own history. Devices of 'meaning production' – grids of visualization, vocabularies, norms, and systems of judgment – *produce* experience; they are not themselves *produced by* experience (cf. Joyce, 1994). These intellectual techniques do not come ready made, but have to be invented, refined, and stabilized, to be disseminated and implanted in different ways in different practices – schools, families, streets, workplaces, courtrooms. If we use the term 'subjectification' to designate all those heterogeneous processes and practices by means of which human beings come to relate to themselves and others as subjects of a certain type, then subjectification has its own history. And the history of subjectification is more practical, more technical, and less unified than sociological accounts allow.

Thus a genealogy of subjectification focuses directly on the *practices* that locate human beings in particular 'regimes of the person'. It does not write a continuous history of the self, but rather accounts for the diversity of languages of 'personhood' that have taken shape – character, personality, identity, reputation, honor, citizen, individual, normal, lunatic, patient, client, husband, mother, daughter – and the norms, techniques, and relations of authority within which these have circulated in legal, domestic, industrial, and other practices for acting upon the conduct of persons. Such an investigation might proceed along a number of linked pathways.

Problematizations

Where, how, and by whom are aspects of the human being rendered problematic, according to what systems of judgment and in relation to what concerns? To take some pertinent examples, one might consider the ways in which the language of constitution and character comes to operate within the themes of urban decline and degeneracy articulated by psychiatrists, urban reformers, and politicians in the last decades of the nineteenth century, or the ways in which the vocabulary of adjustment and maladjustment comes to be used to problematize conduct in sites as diverse as the workplace, the courtroom, and the school in the 1920s and 1930s. To pose the matter in this way is to stress the primacy of the pathological over the normal in the genealogy of subjectification – our vocabularies and techniques of the person, by and large, have not emerged in a field of reflection on the normal individual, the normal character, the normal personality, the normal intelligence, but rather, the very notion of normality has emerged out of a concern with types of conduct, thought, expression deemed troublesome or dangerous (cf. Rose, 1985a). This is a methodological as much as an epistemological point: in the genealogy of subjectification, pride of place is not occupied by the philosophers reflecting in their studies on the nature of the person, the will, the conscience, morality, and the like, but rather in the everyday practices where conduct has become problematic to others or oneself, and in the mundane texts and programs – on asylum management, medical treatment of women, advisable regimes of child rearing, new ideas in workplace management, improving one's self-esteem – seeking to render these problems intelligible and, at the same time, manageable.[4]

Technologies

What means have been invented to govern the human being, to shape or fashion conduct in desired directions, and how have programs sought to embody these in certain technical forms? The notion of technology may seem antithetical to the domain of human being, such that claims about the inappropriate technologization of humanity form the basis of many a critique. However, our very experience of ourselves as certain sorts of persons – creatures of freedom, of liberty, of personal powers, of self-realization – is the outcome of a range of human technologies, technologies that take modes of being human as their object.[5] Technology, here, refers to any assembly structured by a practical rationality governed by a more or less conscious goal. Human technologies are hybrid assemblages of knowledges, instruments, persons, systems of judgment, buildings and spaces, underpinned at the programmatic level by certain presuppositions and objectives about human beings. One can regard the school, the prison, the asylum as examples of one species of such technologies, those which Foucault termed disciplinary and which operate in terms of a detailed structuring of space, time, and relations among individuals, through procedures of hierarchical observation and normalizing judgment, through attempts to enfold these judgments into the procedures and judgments that the individual utilizes in order to conduct his or her own conduct (Foucault, 1977; cf. Markus, 1993, for an examination of the spatial form of such assemblies). A second example of a mobile and multivalent technology is that of the pastoral relation, a relation of spiritual guidance between a figure of authority and each member of his or her flock, embodying techniques such as confession and self-disclosure, exemplarity and discipleship, enfolded into the person through a variety of schema of self-inspection, self-suspicion, self-disclosure, self-decipherment, and self-nurturing. Like discipline, this pastoral technology is capable of articulation in a range of different forms, in the relation of priest and parishioner, therapist and patient, social worker and client, and in the relation of the 'educated' subject to his or her self. We should not see the disciplinary and pastoral relations of subjectification as opposed historically or ethically – the regimes enacted in schools, asylums, and prisons embody both. Perhaps the insistence upon an analytic of human technologies is one of the most distinctive features of the approach I am advocating. Such an analysis does not start from the view that the technologizing of human conduct is malign. Human technologies produce and enframe humans as certain kinds of being whose existence is simultaneously capacitated and governed by their organization within a technological field.

Authorities

Who is accorded or claims the capacity to speak truthfully about humans, their nature and their problems, and what characterizes the truths about persons that are accorded such authority? Through which apparatuses are such authorities authorized – universities, the legal apparatus, churches, politics? To what extent does the authority of authority depend upon a claim to a positive knowledge, to wisdom and virtue, to experience and practical judgment, to the capacity to resolve conflicts? How are authorities themselves governed – by legal codes, by the market, by the protocols of bureaucracy, by professional ethics? And what then is the relation between authorities and those who are subject to them: priest and parishioner, doctor and patient, manager and employee, therapist and client? This focus upon the heterogeneity of authorities, rather than the singularity of 'power' seems to me to be a distinctive feature of genealogies of this sort. They seek to differentiate the diverse persons, things, devices, associations, modes of thought, types of judgment that seek, claim, acquire, or are accorded authority. They chart the different configurations of authority and subjectivity and the varying vectors of force and counterforce installed and made possible. And they seek to explore the variety of ways in which authority has been authorized – not reducing these to the covert intervention of the state or the processes of moral entrepreneurship, but examining in particular the relations between the capacities of authorities and regimes of truth.

Teleologies

What forms of life are the aims, ideals, or exemplars for these different practices for working upon persons: the professional persona exercising a vocation with wisdom and dispassion; the manly warrior pursuing a life of honor through a calculated risking of the body; the responsible father living a life of prudence and moderation; the laborer accepting his or her lot with a docility grounded in a belief in the inviolability of authority or a reward in a life to come; the good wife fulfilling her domestic duties with quiet efficiency and self-effacement; the entrepreneurial individual striving after secular improvements in 'quality of life'; the passionate lover skilled in the arts of pleasure? What codes of knowledge support these ideals, and to what ethical valorization are they tied? Against those who suggest that a single model of the person comes to prominence in any specific culture, it is important to stress the heterogeneity and specificity of the ideals or models of personhood deployed in different practices, and the ways in which they are articulated in relation to specific problems and solutions concerning human conduct. It is only from this perspective, I think, that one can identify the peculiarity of those programmatic attempts to install a single model of the individual as the ethical ideal across a range of different sites and practices. For example, the Puritan sects discussed by Weber were unusual in their attempts to ensure that the mode of individual comportment in terms of sobriety, duty, modesty, self, and so forth applied to practices as diverse as the enjoyment of popular entertainment and labor within the home (cf. Weber, [1905] 1976). In our own times, economics, in the form of a model of economic rationality and rational choice, and psychology, in the form of a model of the psychological individual, have provided the basis for similar attempts at the unification of life conduct around a single model of appropriate subjectivity. But unification of subjectification has to be seen as an objective of particular programs, or a presupposition of particular styles of thinking, not a feature of human cultures.

Strategies

How are these procedures for regulating the capacities of persons linked into wider moral, social, or political objectives concerning the undesirable and desirable features of populations, work force, family, society? Of particular significance here are the divisions and relations established between modalities for the government of conduct accorded the status of political, and those enacted though forms of authority and apparatus deemed nonpolitical – whether these be the technical knowledge of experts, the judicial knowledge of the courts, the organizational knowledge of managers, or the 'natural' knowledges of the family and the mother. Typical of those rationalities of government that consider themselves 'liberal' is the simultaneous delimitation of the sphere of the political by reference to the right of other domains – the market, civil society, the family being the three most commonly deployed – and the invention of a range of techniques that would try to act on events in these domains without breaching their autonomy. It is for this reason that knowledges and forms of expertise concerning the internal characteristics of the domains to be governed assume particular importance in liberal strategies and programs of rule, for these domains are not to be 'dominated' by rule, but must be known, understood, and related to in such a way that events within them – productivity and conditions of trade, the activities of civil associations, ways of rearing children and organizing conjugal relations and financial affairs within household – support, and do not oppose, political objectives.[6] In the case that we are discussing here, the characteristics of persons, as those 'free individuals' upon whom liberalism depends for its political legitimacy and functionality, assume a particular significance. Perhaps one could say that the general strategic field of all those programs of government that regard themselves as liberal has been defined by the problem of how free individuals can be governed such that they enact their freedom appropriately.

The government of others and the government of oneself

Each of these directions for investigation is inspired, in large measure, by the writings of Michel Foucault. In particular, of course, they arise from Foucault's suggestions concerning a genealogy of the arts of government – where government is conceived of, most generally, as encompassing all those more or less rationalized programs and strategies for 'the conduct of conduct' – and his conception of governmentality, which refers to the emergence of political rationalities, or mentalities of rule, where rule becomes a matter of the calculated management of the affairs of each and of all in order to achieve certain desirable objectives (Foucault, 1991; see the discussion of the notion of government in Gordon, 1991). Government, here, does not indicate a theory, but a certain perspective from which one might make intelligible the diversity of attempts by authorities of different sorts to act upon the actions of others in relation to objectives of national prosperity, harmony, virtue, productivity, social order, discipline, emancipation, self-realization, and so forth. This perspective also directs our attention to the ways in which strategies for the conduct of conduct so frequently operate through trying to shape what Foucault termed 'technologies of the self' – 'self-steering mechanisms', or the ways in which individuals experience, understand, judge, and conduct themselves (Foucault, 1986a, 1986b, 1988). Technologies of the self take the form of the elaboration of certain techniques for the conduct of one's relation with oneself, for example, requiring one to relate to oneself epistemologically (know yourself), despotically (master yourself), or in other ways (care for yourself). They are embodied in particular technical practices (confession, diary writing, group discussion, the twelve-step program of Alcoholics Anonymous). And they are always practiced under the actual or imagined authority of some system of truth and of some authoritative individual, whether this be theological and priestly, psychological and therapeutic, or disciplinary and tutelary.

A number of issues arise from these considerations.

The first concerns the issue of ethics itself. In his later writings, Foucault utilized the notion of 'ethics' as a general designation for his investigations into the genealogy of our present forms of 'concern' for the self (Foucault, 1979b, 1986a, 1986b; cf. Minson, 1993). Ethical practices for Foucault were distinguished from the domain of morality, in that moral systems are, by and large, universal systems of injunction and interdiction – thou shalt do this or thou shalt not do that – and are most frequently articulated in relation to some relatively formalized code. Ethics, on the other hand, refers to the domain of specific types of practical advice as to how one should concern oneself with oneself, make oneself the subject of solicitude and attention, conduct oneself in the various aspects of one's everyday existence. Different cultural periods, Foucault argued, differed in the respective weight that their practices for the regulation of conduct placed upon codified moral injunctions and the practical repertoires of ethical advice. However, one might undertake a genealogy of our contemporary ethical regime, which, Foucault suggested, encouraged human beings to relate to themselves as the subject of a 'sexuality', and to 'know themselves' through a hermeneutics of the self, to explore, discover, reveal, and live in the light of the desires that composed their truth. Such a genealogy would disturb the appearance of enlightenment that clothed such a regime, by exploring the way in which certain forms of spiritual practice that could be found in Greek, Roman, and Early Christian ethics had become incorporated into priestly power, and later into the practices of the educational, medical, and psychological type (Foucault, 1986b, p. 11).

Clearly the approach I have outlined has derived much from Foucault's way of thinking about these issues. However, I would wish to develop his arguments in a number of respects. First, as has been pointed out elsewhere, the notion of 'techniques of the self' can be somewhat misleading. The self does not form the transhistorical object of techniques for being human but only one way in which humans have been enjoined to understand and relate to themselves (Hadot, 1992). In different practices, these relations are cast in terms of individuality, character, constitution, reputation, personality, and the like, which neither are merely *different versions of* a self, nor *sum into* a self. Further, the extent to which our contemporary relation to ourselves – inwardness, self-exploration, self-fulfillment, and the like – does indeed take the issue of sexuality and desire as its fulcrum must remain an open question for historical investigation. Elsewhere I have suggested that the self, itself, has become the object of valorization, a regime of subjectification in which desire has been freed from its dependence upon the law of an inner sexuality and been transformed into a variety of passions to discover and realize the identity of the self itself (Rose, 1990).

Further, I would suggest, one needs to extend an analysis of the relations between government and subjectification beyond the field of ethics, if by that one means all those styles of relating to oneself that are structured by the divisions of truth and falsity, the permitted and the forbidden. One needs to examine, also, the government of this relation along some other axes.

One of these axes concerns the attempt to inculcate a certain relation to oneself through transformations in 'mentalities' or what one might term 'intellectual techniques' – reading, memory, writing, numeracy, and so forth (see, for some powerful examples, Eisenstein, 1979, and Goody and Watt, 1963). For example, especially over the course of the nineteenth century in Europe and the United States, one sees the development of a host of projects for the transformation of the intellect in the service of particular objectives, each of which seeks to enjoin a particular relation to the self through the implantation of certain capacities of reading, writing, and calculating. One example here would be the way in which, in the latter decades of the nineteenth century, Republican educators in the United States promoted numeracy, in particular the numerical capacities that would be facilitated by decimalization, in order to generate a particular kind of relation to themselves and their world in those so equipped. A numerate self would be a calculating self, who would establish a prudent relation to the future, to budgeting, to trade, to politics, and to life conduct in general (Cline-Cohen, 1982, pp. 148–9; cf. Rose, 1991).

A second axis would concern corporealities or body techniques. Of course, anthropologists and others have investigated in detail the cultural shaping of bodies – comportment, expression of emotion, and the like – as they differ from culture to culture, and within cultures between genders, ages, status groups, and so forth. Marcel Mauss provides the classic account of the ways in which the body, as a technical instrument, is organized differently in different cultures – different ways of walking, sitting, digging, marching (Mauss, 1979a; cf. Bourdieu, 1977). However, a genealogy of subjectification is not concerned with the cultural relativity of bodily capacities in and of itself, but with the ways in which different corporeal regimes have been devised and implanted in rationalized attempts to produce a particular relation to the self and to others. Norbert Elias has given many powerful examples of the ways in which explicit codes of bodily conduct – manners, etiquette, and the self-monitoring of bodily functions and actions – were enjoined upon individuals in different positions within the apparatus of the court of Louis XIV in the mid-eighteenth century (Elias, 1983; cf. Elias, 1978; Osborne, 1996). The disciplining of the body of the pathological individual in the nineteenth-century prison and the asylum did not only involve its organization within an external regime of hierarchical surveillance and normalizing judgment and its assembling through molecular regimes governing movement in time and space: it also sought to enjoin an internal relation between the pathological individual and his or her body, in which bodily comportment would both manifest and maintain a certain disciplined mastery exercised by the person over himself or herself (Foucault, 1967, 1977; see also Smith, 1992, for a history of the notion of 'inhibition' and its relation to the Victorian concern with the outward manifestation of steadfastness and self-mastery through the exercise of control over the body). An analogous, though substantively very different, relation to the body was a key element in the self-sculpting of a certain aesthetic persona in nineteenth-century Europe, embodied in certain styles of dress but also in the cultivation of certain body techniques such as swimming that would produce and display a particular relation to the natural (Sprawson, 1992). Theorists of gender have begun to analyze the ways in which the appropriate performance of sexual identity has historically been linked to the inculcation of certain techniques of the body (Brown, 1989; Butler, 1990; Bordo, 1993) Certain ways of holding oneself, walking, running, holding the head, and positioning the limbs are not merely culturally relative or acquired through gender socialization, but are regimes of the body that seek to subjectify in terms of a certain truth of gender, inscribing a particular relation to oneself in a corporeal regime: prescribed, rationalized, and taught in manuals of advice, etiquette, and manners, and enjoined by sanctions as well as seductions (cf. the studies collected in Bremer and Roodenburg, 1991).

These comments should indicate something of the heterogeneity of the links between the government of others and the government of the self. It is important to stress two further aspects of this heterogeneity. The first concerns the diversity of modes in which a certain relation to oneself is enjoined. There is a temptation to stress the elements of self-mastery and restrictions over one's desires and instincts that are entailed in many regimes of subjectification – the injunction to control or civilize an inner nature that is excessive. Certainly one can see this theme in many nineteenth-century debates over ethics and character for both the ruling orders and in the respectable laboring classes – a paradoxical 'despotism of the self' at the heart of liberal doctrines of liberty of the subject (I derive this formulation from Valverde, 1996; cf. Valverde, 1991). However there are many other modes in which this relation to oneself can be established and, even within the exercise of mastery, a variety of configurations through which one can be encouraged to master oneself (cf. Sedgwick, 1993). To master one's will in the service of character through the inculcation of habits and rituals of self-denial, prudence, and foresight, for example, is different from mastering one's desire through bringing its roots to awareness through a reflexive hermeneutics in order to free oneself from the self-destructive consequences of repression, projection, and identification.

Further, the very form of the relation can vary. It can be one of knowledge, as in the injunction to know oneself, which Foucault traces back to the Christian confession and forward to contemporary techniques of psychotherapeutics: here the codes of knowledge are inevitably supplied not by pure introspection but by rendering one's introspection in a particular vocabulary of feelings, beliefs, passions, desires, values, or whatever and according to a particular explanatory code derived from some source of authority. Or it can be one of concern and solicitude, as in projects for the care of the self through acting upon the body which is to be nurtured, protected, safeguarded by regimes of diet, stress minimization, and self-esteem. Equally, the relation to authority can vary. Consider, for example, some of the changing authority configurations in the government of madness and mental health: the relation of mastery that was exercised between asylum doctor and mad person in late eighteenth-century moral medicine; the relation of discipline and institutional authority that obtained between the nineteenth-century asylum doctor and the inmate; the relation of pedagogy that obtained between the mental hygienists of the first half of the twentieth century and the children and parents, pupils and teachers, workers and managers, generals and soldiers upon whom they sought to act; the relation of seduction, conversion, and exemplarity that obtains between the psychotherapist and the client today.

As will be evident from the preceding discussion, although the relations to oneself enjoined at any one historical moment may resemble one another in various ways – for example the Victorian notion of character was widely dispersed across many different practices – it is for empirical investigation to map the topography of subjectification. It is not a matter, therefore, of narrating a general history of the idea of the person or self, but of tracing the technical forms accorded to the relation to oneself in various practices – legal, military, industrial, familial, economic. And even within any practice, heterogeneity must be assumed to be more common than homogeneity – consider, for example, the very different configurations of personhood in the legal apparatus at any one moment, the difference between the notion of status and reputation as it functioned in civil proceedings in the nineteenth century and the simultaneous elaboration of a new relation to the lawbreaker as a pathological personality in the criminal courts and the prison system (cf. Pasquino, 1991).

Our own present certainly appears to be marked by a certain leveling of these differences, such that presuppositions concerning human beings in diverse practices share a certain family resemblance – humans as selves with autonomy, choice, and self-responsibility, equipped with a psychology aspiring to self-fulfillment, actually or potentially running their lives as a kind of enterprise of themselves. But this is precisely the point of departure for a genealogical investigation. In what ways was this regime of the self put together, under what conditions and in relation to what demands and forms of authority? We have undoubtedly seen a proliferation of expertises of human conduct over the past hundred years: economists, managers, accountants, lawyers, counselors, therapists, medics, anthropologists, political scientists, social policy experts, and the like. But I would argue that the 'unification' of regimes of subjectification in terms of the self has much to do with the rise of one particular form of positive expertise of human being – that of the psy disciplines, and their 'generosity'. By their 'generosity' I mean that, contrary to conventional views of the exclusivity of professional knowledge, psy has been happy, indeed eager, to 'give itself away' – to lend its vocabularies, explanations, and types of judgment to other professional groups and to implant them within its clients (Rose, 1992b; see this volume, Chapter 4). The psy disciplines, partly as a consequence of their heterogeneity and lack of a single paradigm, have acquired a peculiar penetrative capacity in relation to practices for the conduct of conduct. They have not only been able to supply a whole variety of models of selfhood but also to provide practicable recipes for action in relation to the government of persons by professionals in different locales. Their potency has been further increased by their ability to supplement these practicable qualities with a legitimacy deriving from their claims to tell the truth about human beings. They have disseminated themselves rapidly through their ready translatability into programs for reshaping the self-steering mechanisms of individuals, whether these be in the clinic, the classroom, the consulting room, the magazine advice column, or the confessional television show. It is, of course, true that the psy disciplines are not held publicly in particularly high esteem, and their practitioners are often figures of fun. But one should not be misled by this – it has become impossible to conceive of personhood, to experience one's own or another's personhood, or to govern oneself or others without psy.

Let me return to the issue of the diversity of regimes of subjectification. A further dimension of heterogeneity arises from the fact that ways of governing others are linked not only to the subjectification of the governed, but also to the subjectification of those who would govern conduct. Thus Foucault argues that the problematization of sex between men for the Greeks was linked to the demand that one who would exercise authority over others should first be able to exercise dominion over his own passions and appetites – for only if one was not a slave to oneself was one competent to exercise authority over others (Foucault, 1988; cf. Minson, 1993, pp. 20–1). Peter Brown points to the work required of a young man of the privileged classes in the Roman Empire of the second century, who was advised to remove from himself all aspects of 'softness' and 'womanishness' – in his gait, in his rhythms of speech, in his self-control – in order to manifest himself as capable of exercising authority over others (Brown, 1989, p. 11). Gerhard Oestreich suggests that the revival of Stoic ethics in seventeenth- and eighteenth-century Europe was a response to the criticism of authority as ossified and corrupt: the virtues of love, trust, reputation, gentleness, spiritual powers, respect for justice, and the like were to become the means for authorities to renew themselves (Oestreich, 1982, p. 87). Stephan Collini has described the novel ways in which the Victorian intellectual classes problematized themselves in terms of such qualities as steadfastness and altruism: they interrogated themselves in terms of a constant anxiety about and infirmity of the will, and found, in certain forms of social and philanthropic work, an antidote to self-doubt (Collini, 1991, discussed in Osborne, 1996). While these same Victorian intellectuals were problematizing all sorts of aspects of social life in terms of moral character, threats to character, weakness of character, and the need to promote good character, and arguing that the virtues of character – self-reliance, sobriety, independence, self-restraint, respectability, self-improvement – should be inculcated in others through positive actions of the state and the statesman, they were making themselves the subject of a related, but rather different, ethical work (Collini, 1979, pp. 29–32). Similarly, throughout the nineteenth century, one sees the emergence of quite novel programs for the reform of secular authority within the civil service, the apparatus of colonial rule, and the organizations of industry and politics, in which the persona of the civil servant, the bureaucrat, the colonial governor will become the target of a whole new ethical regime of disinterest, justice, respect for rules, distinction between the performance of one's office and one's private passions, and much more (Weber, 1978: cf. Hunter, 1993a, b, c; Minson, 1993; du Gay, 1995; Osborne, 1994). And, of course, many of those who were subject to the government of these authorities – indigenous officials in the colonies, housewives of the respectable classes, parents, schoolteachers, working men, governesses – were themselves called upon to play their part in the making up of persons and to inculcate in them a certain relation to themselves.

From this perspective, it is no longer surprising that human beings often find themselves resisting the forms of personhood that they are enjoined to adopt. Resistance – if by that one means opposition to a particular regime for the conduct of one's conduct – requires no theory of agency. It needs no account of the inherent forces within each human being that love liberty, seek to enhance their own powers or capacities, or strive for emancipation, that are prior to and in conflict with the demands of civilization and discipline. One no more needs a theory of agency to account for resistance than one needs an epistemology to account for the production of truth effects. Human beings are not the unified subjects of some coherent regime of government that produces persons in the form in which it dreams. On the contrary, they live their lives in a constant movement across different practices that subjectify them in different ways. Within these different practices, persons are addressed as different sorts of human being, presupposed to be different sorts of human being, acted upon as if they were different sorts of human being. Techniques of relating to oneself as a subject of unique capacities worthy of respect run up against practices of relating to oneself as the target of discipline, duty, and docility. The humanist demand that one deciphers oneself in terms of the authenticity of one's actions runs up against the political or institutional demand that one abides by the collective responsibility of organizational decision making even when one is personally opposed to it. The ethical demand to suffer one's sorrows in silence and find a way of 'going on' is deemed problematic from the perspective of a passional ethic that obligates one to disclose oneself in terms of a particular vocabulary of emotions and feelings.

Thus the existence of contestation, conflict, and opposition in practices that conduct the conduct of persons is no surprise and requires no appeal to the particular qualities of human agency, except in the minimal sense that human being – like all else – exceeds all attempts to think it; while human being is necessarily thought, it does not exist in the form of thought.[7] Thus, in any one site or locale, humans turn programs intended for one end to the service of others. For example, psychologists, management reformers, unions, and workers have turned to account the vocabulary of humanistic psychology in a criticism of practices of management based on a psychophysiological or disciplinary understanding of persons. Reformers of the practices of welfare and medicine have, over the past two decades, turned to the notion that human beings are subjects of rights against practices that presuppose human beings as the subjects of care. Out of this complex and contested field of oppositions, alliances, and disparities of regimes of subjectification come accusations of inhumanity, criticisms, demands for reform, alternative programs, and the invention of new regimes of subjectification.

If we choose to designate some dimensions of these conflicts resistance, this itself is perspectival: it requires us to exercise a judgment. It is fruitless to complain that such a perspective gives one no place to stand in the making of ethical critique and in the evaluation of ethical positions. The history of all those attempts to ground ethics that do not appeal to some transcendental guarantor is plain enough – they cannot close conflicts over regimes of the person, but simply occupy one more position within the field of contestation (MacIntyre, 1981).

Folds in the soul

But are not the kind of phenomena I have been discussing of interest precisely *because* they produce us as human beings with a certain kind of subjectivity? This is certainly the view of many who have investigated these issues, from Norbert Elias to contemporary feminist theorists who rely upon psychoanalysis to ground an account of the ways in which certain practices of the self become inscribed within the body and soul of the gendered subject (e.g., Butler, 1993; Probyn, 1993). For some, this path appears unproblematic. Elias, for example, did not doubt that human beings were the type of creatures inhabited by a psychoanalytic psychodynamics, and that this provided the material basis for the inscription of civility into the soul of the social subject (Elias, 1978). I have already suggested that such a view is paradoxical, for it requires us to adopt a recent historical truth about the human being – that carved out at the end of the nineteenth century – as the universal basis for investigating the historicity of being human. For others, such a choice is required if one is to avoid representing the human being as merely the passive and interminably malleable object of historical processes, if one is to have an account of agency and of resistance, and if one is to be able to find a place to stand in order to evaluate one regime of personhood over and above another (for one example of this argument, see Fraser, 1989). I have suggested that no such theory is required to account for conflict and contestation, and the stable ethical ground apparently provided by any given theory of the nature of human beings is illusory. One has no choice but to enter into a debate that cannot be closed by appeal to the nature of the human being as essentially and universally a subject of rights, of freedom, of autonomy, or whatever. Is it possible, then, that one might write a genealogy of subjectification without a metapsychology? I think it is.

Such a genealogy, I suggest, requires only a minimal, weak, or thin conception of the human material on which history writes (cf. Patton, 1994). We are not concerned here with the social or historical construction of the person or with the narration of the birth of modern self-identity. Our concern, rather, is with the diversity of strategies and tactics of subjectification that have taken place and been deployed in diverse practices at different moments and in relation to different classifications and differentiations of persons. The human being, here, is not an entity with a history, but the target of a multiplicity of types of work, more like a latitude or a longitude at which different vectors of different speeds intersect. The 'interiority' which so many feel compelled to diagnose is not that of a psychological system, but of a discontinuous surface, a kind of infolding of exteriority.

I draw this notion of folding loosely from the work of Gilles Deleuze (Deleuze, 1988, 1990a, 1992a; cf. Probyn, 1993, pp. 128–34). The concept of the fold or the pleat suggests a way in which we might think of an internality being brought into existence in the human being without postulating any prior interiority, and thus without binding ourselves to a particular version of the law of this interiority whose history we are seeking to diagnose and disturb. The fold indicates a relation without an essential interior, one in which what is 'inside' is merely an infolding of an exterior. We are familiar with the idea that aspects of the body which we commonly think of as part of its interiority – the digestive tract, the lungs – are no more than the invagination of an outside. This does not prevent them from being invested with both personal and cultural affects and values in terms of an apparently immutable body image that is taken as the norm for our perception of the contours and limits of our corporeality. Perhaps, then, we might think of the power that modes of subjectification have upon human beings in terms of such an infolding. Folds incorporate without totalizing, internalize without unifying, collect together discontinuously in the form of pleats making surfaces, spaces, flows, and relations.

Within a genealogy of subjectification, that which would be infolded would be anything that can acquire authority: injunctions, advice, techniques, little habits of thought and emotion, an array of routines and norms of being human – the instruments through which being constitutes itself in different practices and relations. These infoldings are partially stabilized to the extent that human beings have come to imagine themselves as the subjects of a biography, to utilize certain 'arts of memory' in order to render this biography stable, to employ certain vocabularies and explanations to make this intelligible to themselves. This is indicative of the need to extend the limits of the metaphor of the fold. For the lines of these folds do not run through a domain coterminous with the fleshly bounds of the human epidermis. Human being is emplaced, enacted through a regime of devices, gazes, techniques that extend beyond the limits of the flesh. Memory of one's biography is not a simple psychological capacity, but is organized through rituals of storytelling, supported by artifacts such as photograph albums and so forth. The regimes of bureaucracy are not merely ethical procedures infolded into the soul, but occupy a matrix of offices, files, typewriters, habits of timekeeping, conversational repertoires, techniques of notation. The regimes of passion are not merely affective folds in the soul, but are enacted in certain secluded or valorized spaces, through sensualized equipment of beds, drapes, and silks, routines of dressing and undressing, aestheticized devices for providing music and light, regimes of partitioning of time, and so forth (cf. Ranum, 1989). Folding being is not a matter of bodies, but of assembled locales.

We might counterpose such a *spatialization* of human being to the narrativization undertaken by sociologists and philosophers of modernity and postmodernity. That is to say, we need to render human being intelligible in terms of assemblages (I develop this argument in Chapter 8). By assemblages I mean the localization and connecting together of routines, habits, and techniques within specific domains of action and value: libraries and studies, bedrooms and bathhouses, courtrooms and schoolrooms, consulting rooms and museum galleries, markets and department stores. The five volumes of *The History of Private Life* compiled under the general editorship of Phillipe Ariès and George Duby provide a wealth of examples of the way in which novel human capacities such as styles of writing or sexuality depend upon and give rise to particular forms of spatial organization of the human habitat (Veyne, 1987; Duby, 1988; Chartier, 1989; Perrot, 1990; Prost and Vincent, 1991). However, there is nothing privileged about what has become termed 'private life' for the emplacement of regimes of subjectification – it is in the factory as much as the kitchen, in the military as much as the study, in the office as much as the bedroom, that the modern subject has been required to identify his or her subjectivity. To the apparent linearity, unidirectionality, and irreversibility of time, we can counterpose the multiplicity of places, planes, and practices. And in each of these assemblages, repertoires of conduct are activated that are not bounded by the enclosure formed by the human skin or carried in a stable form in the interior of an individual: they are rather webs of tension across a space that accord human beings capacities and powers to the extent that they catch them up in hybrid assemblages of knowledges, instruments, vocabularies, systems of judgment, and technical devices. To this extent a genealogy of subjectification needs to think human being as a kind of machination, a hybrid of flesh, artifact, knowledge, passion, and technique.

Conclusion

It is characteristic of our current regime of the self to reflect on and act on all the diverse domains, practices, and assemblages in terms of a unified 'personality', an 'identity' to be revealed, discovered, or worked on in each. This machination of the self in terms of identity needs to be recognized as a regime of subjectification of recent origin. In the essays that follow, I argue that the psy disciplines have played a key role in our contemporary regime of subjectification, and its unification under the sign of the self. Thus a critical history of psy would take as its object our contemporary regime of the self and its identity – together with all the judgments and judges that have populated it. It would describe the part played by the psychosciences in the genealogy of that regime, and the relations it constructs between the one and the many, the internal and the external, the whole and the part in the classifications that have been carved out within its volume. A genealogy of psychology's contribution to our regime of self thus connects up, in a lateral fashion, with all those contemporary political movements that have challenged the category of identity – the identity of woman, the identity of race, the identity of class (see, in particular, Haraway, 1991, and Riley, 1988). If one leaves aside the frothy 'postmodern' celebrations of the playfulness of 'difference', such challenges are motivated, in part, by the belief that values of self and identity are not so much *resources* for critical thought as *obstacles* to such thought. The politics of identity, even when not associated with barbarous projects for the 'cleansing' of difference, are racked by internal fragmentations in which the subjects supposed to be unified – as women, as black, as disabled, as mad – refuse to recognize themselves in the name that is offered to them. In this fragmentation and these refusals, we have been forced to recognize that national, racial, sexual, gendered, class identities have, historically, been created most typically by those who would identify us in the service of problematizing, regulating, policing, reforming, improving, developing, or even eliminating those so identified. Of course, such identities have often been embraced by those so identified and turned back upon the regimes that have created them. But to declare "I *am* that name": woman, homosexual, proletarian, African American – or even man, white, civilized, responsible, masculine – is no outward representation of an inward and spiritual state but a response to that history of identification and its ambiguous gifts and legacies.

It is true that we cannot analyze the present by reference to the sins that may lie in its genealogies. The vocabularies that we utilize to think ourselves arise out of our history, but do not always bear upon them the marks of their birth: the historicity of concepts is too contingent, too mobile, opportunistic, and innovative for this. The political strategies motivated by the ideals of identity were, no doubt, imbued as often by the noble values of humanism and its commitment to individual liberty as by a will to dominate or purify in the name of identity. But as our own century ends, perhaps it is time to try to count the costs, and not just the blessings, of our identity projects. And one small but significant element of counting those costs lies in identifying the contributions made to such a regime of the subjectification by psychology, as the discourse that for some hundred fifty years has spoken to us – sometimes in brutal commands, sometimes in dispassionate disquisitions, sometimes in seductive and comforting whispers – of the truths of ourselves.

2

A critical history of psychology

How should one do the history of psychology? I would like to propose a particular approach to this issue, a critical history of the relations between the psychological, the governmental, and the subjective. A critical history is one that helps us think about our nature and our limits, about the conditions under which that which we take for truth and reality has been established. Critical history disturbs and fragments, it reveals the fragility of that which seems solid, the contingency of that which seemed necessary, the mundane and quotidian roots of that which claims lofty nobility. It enables us to think *against* the present, in the sense of exploring its horizons and its conditions of possibility. Its aim is not to predetermine judgment, but to make judgment possible.

Psychology and its histories

The psychological sciences – psychology, psychiatry, and the other disciplines that designate themselves with the prefix 'psy' – are certainly not devoid of a historical consciousness. Many weighty tomes tell the story of the long development of the scientific study of psychological functioning, normal and pathological. Almost every psychiatric or psychological textbook appears obliged to include a historical chapter or review, however desultory, of the topics under discussion. These texts repeatedly tell us the story of the development of the psychological sciences in similar terms: they have a long past but a short history. A long past: a continuous tradition of speculation concerning the nature, vicissitudes, and pathologies of the human soul, virtually coextensive with the human intellect itself. A short history: the shift from metaphysics, speculation, or medical or physiological reduction, which occurs only with the deployment of 'the experimental method' in the nineteenth century. It is tempting to dismiss these histories for their epistemological naiveté, or to see peculiarly self-interested motives at work among those writing the story of the sciences of mind. Each of these accusations may contain some truth. But this way of using history is not unique to the psychological sciences. Indeed a certain form of history is an internal element in the consciousness of all those practices of representing and intervening which we call science.

Such authoritative texts of scientific history play a key role in constructing the image of the present reality of the discipline in question, a role that is indicated by the part they play in the training of every novitiate. Georges Canguilhem terms such writing 'recurrent history' (Canguilhem, 1968, 1977). He uses this term to describe – not necessarily pejoratively – the ways in which scientific disciplines tend to identify themselves partly through a certain conception of their past. These narratives establish the unity of the science by constructing a continuous tradition of thinkers who sought to grasp the phenomena that form its subject matter. Inescapably, from such a perspective, the object of a science – the 'reality' it seeks to render intelligible – appears both ahistorical and asocial. It preexists the attempts to study it, it has always existed in the same form, and all these thinkers of the past have circled around a reality that has remained the same. Thus the works of these thinkers can be ordered into a narrative, arranged along a chronological dimension, which corresponds to a progress toward the object; disruption to this smooth progression can be reincorporated into the narrative via the notions of precursor, genius, prejudice, and influence.

Simultaneously these recurrent histories establish the modernity of their science. They both ratify the present through its respectable tradition, and demarcate it from those aspects of the past that might disturb it. They do this by effecting a division between sanctioned and lapsed texts and authors, between those theories and arguments which are of a piece with the current self-image of the discipline and those which are marginal and eccentric. The sanctioned past is arranged in a more or less continual sequence, as that which led to the present and anticipated it, that virtuous tradition of which the present is the inheritor. It is a past of individual insights, difficult advances, and unexpected failures, of personal, professional, and cultural influences, of obstacles overcome, crucial experiments, original discoveries, and the like. Opposed to this sanctioned history is a lapsed history. This is a history of false paths, of errors and illusions, of prejudice and mystification – all those cul-de-sacs into which knowledge was drawn and which diverted it from the path of progress. Consigned to this history of error are all those books, theories, arguments, and explanations associated with the past of a system of thought but incongruous with its present. Recurrent histories take the present as both the culmination of the past and the standpoint from which its historicity can be displayed. Recurrent histories are, however, more than 'ideology'; they have a constitutive role to play in most scientific discourses. For they use the past to help demarcate that regime of truth which is contemporary for a discipline – and in doing so, they not only use history to police the present, but also to shape the future (the most discussed example is, of course, Boring, 1929). Such histories police the boundaries of their discipline by their criteria of inclusion or exclusion. Hence such histories play their part in establishing a division between the sayable and the unsayable, the thinkable and the unthinkable: in emplacing what Michel Foucault has termed a 'regime of truth'.

These recurrent histories of science are programmatic. In narrating the past of their discipline they seek not only to demarcate the present but to write the future. They thus write their histories in the future anterior. Now I would also like to urge us to do 'present centered' history. But such a 'history of the present' has to take the current image of the discipline as both a claim and a problem: a claim in that we need to examine this image neither as myth nor as reflection but examine how it operates and the functions it performs within the discipline today; and a problem in that we cannot ourselves use it as the basis for our investigation of the past. What today appears marginal, eccentric, or disreputable was frequently, at the time it was written, central, normal, and respectable. Rather than marginalize these texts of the past from the point of view of the present, we might do better to question the certainties of the present by attention to such margins and to the process of their marginalization. And indeed, when analyzed in this way, the apparent certainties of our present disciplinary identities also begin to break up. Not only do disciplines have fluid boundaries, one with another, but the lines of development of theory, explanation, and experimentation often do not run through the center of any one discipline but across its links with others, by means of questions that have less to do with knowledge than with know-how. Such a critical history of the present should be a disturbing, disrupting, and fragmenting operation.

Until the 1960s, almost all histories of psychology were of the 'recurrent' genre (a situation described and criticized by Young, 1966). In the subsequent period, however, this recurrent history of the psychological sciences has been challenged on a number of fronts. Sociologists of social control and cultural critics have extended their critique to psychology. A new 'social' history of science has gone beyond the classical division between internal and external histories of science and argued, in a range of different ways, that scientific knowledge itself must be understood in its social, political, institutional context and in terms of the organization of scientific communities. Further, there has been an upsurge in scholarly histories of the psychosciences, which have examined in great detail the biographical and institutional forces that shaped the development of psychological theories and techniques, the organizational forces at work within the academic world, the political influences on the development of psychological knowledge (for some representative collections, see Woodward and Ash, 1982; Ash and Woodward, 1987). I do not wish to discuss these different approaches in any great detail. At the risk of caricature, however, I may be able to clarify the project that I have termed a critical history of psychology by contrasting it with these other perspectives.

Sociological critiques that have touched on the psychological sciences have sought to oppose and revise the themes of progress, enlightenment, and neutrality that animate authoritative history, characterizing these works as self-serving hagiographies whose aim is not to enlighten us about the past but to legitimize the present. They oppose this programmatics of legitimation with a politics of delegitimation. They analyze the development of the disciplines less in terms of the innovative power of genius or the corrective power of scientific experimentation than in terms of transformations external to scientific knowledge. As far as the nineteenth and twentieth centuries are concerned, these analyses tend to place more or less weight on five different sorts of external factors: economic, professional, political, cultural, and patriarchal. Economic themes link up developments in the psychological sciences in the nineteenth century with the exigencies of capitalist production, the construction and regulation of the labor market, and the preservation of the property and authority of the wealthy, and, more recently, with the colonial adventures of domination and plunder to which the rise of metropolitan capitalism was inherently bound. Professional themes link up the formulation and adoption of different theories, explanations, and techniques with the clash of cognitive and professional interests, sometimes analyzed in terms of class, and with the extension of professional power by means of the authority deriving from a successful claim to science. Political themes link these developments with transformations in the apparatus of the state and the institutions of social control such as the asylum and the prison. Cultural themes tend to see the rise of psychology as an instance of a broader social malaise: the decline of spiritual and communal values, the revised relations of public and private and the tyranny of intimacy, the rise of narcissism at the level of individuals and cultures. Patriarchal themes have linked the rise of the psychosciences with the nineteenth-century domestication of women and the sequestering of wives and daughters in the claustrophobic and pathogenic confines of the nuclear family.

Such writing of history as critique poses significant questions concerning the relations between knowledge and society, between truth and power, between psychology and subjectivity. However, this use of history is often as problematic as the authoritative versions it contests. I suggest that an effective critical history would need to reverse our direction of investigation in relation to each of these themes.

Economic factors

Explanations that rest on economic exigencies are rarely able to specify exactly the mechanisms whereby economic developments were translated into specific shifts in knowledge except by resorting to the feeble explanatory powers of notions such as legitimation (see, e.g., Baritz, 1960; Ewen, 1976, 1988). I suggest that we might throw more light on the relation between the vicissitudes of capitalism and the rise of the psychological disciplines by examining, instead, the political, institutional, and conceptual conditions that gave rise to the formulation of different notions of the economy, the market, the laboring classes, the colonial subject. We should attend to the ways in which these problematized different aspects of existence (industrial disruption, productivity, health of the laborer whether free or slave, effective management of colonial plantations) from the perspective of 'the economy'. We should analyze the ways in which these problematizations produced questions to which the psychosciences could come to provide answers. And we should explore the ways in which the psychosciences, in their turn, transformed the very nature and meaning of economic life and the conceptions of economic exigencies that have been adopted in economic activity and policy.

Professional factors

A similar reversal could be adopted in relation to the question of interests. Sociologists have apparently regarded it as a simple matter to ascribe interests to individuals or groupings – classes, genders, races – and utilize these as explanations of the positions taken up in cognitive or professional disputes (this is especially true of the Edinburgh group of sociologists of science: Barnes, 1974; Mackenzie, 1981). Unfortunately, these explanations frequently lapse into the tautological. Usually the interest is derived from the position adopted, which it then purports to explain: because some psychologists elaborated the view that women's mental capacities were linked to their reproductive cycles, they must have had an interest in portraying women as unstable and hence dependent, and therefore this interest explains why they thought as they did. Alternatively the relation between the interest and the point of view is established by hindsight – naturally, (male) psychologists after the end of World War II would have come to the view that there was a 'maternal instinct' because, after all, there was a 'need' to return women from factory to home in the 1950s (such arguments are critically examined in Riley, 1983, pp. 109–49). These explanations simply assume what they set out to explain. Instead, I suggest, we need to account for the very formation of interests themselves. We need to address ourselves to the ways in which specific individuals and groups were mobilized around particular objectives, to the techniques for construction of collective identities and aspirations. From this perspective, *claims* about what is in whose interests, to the extent that they produce allies, actually constitute the groups, communities, forces in question, whether these be industrialists, male laborers in manufacturing, bourgeois women, or psychological professionals. We thus need to study the way in which alliances come to be formed between those who come to be convinced, in different ways, that they have certain interests and that their interests are the same as those of particular others (cf. Callon, 1986; Latour, 1984, 1986a). Interests are *achievements,* not *explanations,* and are more fragile, more contested and more negotiated and negotiable than many sociologists and others like to believe.

Political factors

Sociological histories of the psychological sciences often see 'the state' as the origin, orchestrator, or beneficiary of many of the social practices carried out in the name of psychology or psychiatry (see, e.g., the essays collected in Cohen and Scull, 1983). Again, I would like to turn this problem on its head. It is precisely the birth of this conception of the state that should be investigated. Rather than investigating the state extending its control in the nineteenth century, and the psychological sciences serving such functions for it, we should investigate the formation of a new way of mobilizing political authority in this period. Although the theme of the 'nineteenth-century revolution in government' is familiar, less noticed is the role played by the psychological sciences in the birth of this new form of governmental rationality that entails a new understanding of the state, and a new way of constituting the population of a particular national territory as political subjects (Rose and Miller, 1992; cf. MacDonagh, 1958, 1977; MacLeod, 1988). The disciplinization of psychology is constitutively bound to a fundamental transformation that has occurred in the rationalities and technologies of political power since the last decades of the nineteenth century, in which the responsibility of rulers has come to be posed in terms of securing the welfare and normality, physical and mental, of citizens, and of shaping and regulating the ways in which they conduct their 'private' existence – as workers, citizens, fathers, mothers – such that they enact their privacy and freedom according to these norms of maximized normality. The field of power that is codified as the state is intelligible only when located within this wider matrix of projects, programs, and strategies for the conduct of conduct, elaborated and enacted by a whole diversity of authorities shaping and contesting the very boundaries of the political (Foucault, 1991).

Cultural factors

Cultural critics have tended to view the rise of psychology in the twentieth century as merely a symptom of the mentality of an age that has seen the birth of the inward-looking, isolated, self-sufficient individual, for whom truth is neither collective, nor sacred, but personal (Rieff, 1966; Sennett, 1977; Lasch, 1979). But, again, the direction of investigation might be reversed, to focus less on the 'mentalities' that give rise to ethics, and more on the specific conditions of emergence, articulation, and transformation of the ethical values and techniques that make certain cultural practices possible. From such a perspective, the question to be addressed in a critical history of psychology concerns the ways in which, at different historical moments and in relation to different problems and persons, ethical practices have drawn on aspects of psy knowledge, technical procedures, and authoritative persons in acting upon the self-steering mechanisms of individuals. Psychology here would not be viewed in terms of cultural beliefs and meanings, but would take its place within a genealogy of 'technologies of subjectification': the practical rationalities that human beings have applied to themselves and others in the name of self-discipline, self-mastery, beauty, grace, virtue, or happiness.

Patriarchal factors

Perhaps the most powerful recent historical critiques of the psychosciences have been written by feminists, seeking to document the part played by psychology and psychiatry over the second half of the nineteenth century and the whole of the twentieth in promulgating a certain myth of woman that would shore up a patriarchal order, and legitimate the infantilization of women, the reproduction of their dependency, and their subordination in domestic relations, the private world of the home and the burden of motherhood in the name of their delicacy, their psychological vulnerability, and their natural maternality (Showalter, 1987; Ussher, 1991; Badinter, 1981). Such work has played a key role in critical thought, in particular, in examining the extent to which identities and attributes of males and females that have conventionally been placed on the side of the natural have been constructed around a whole variety of problems of regulation, in relation to a whole variety of cultural assumptions and practices for the administration of space (e.g., public and private) and interaction (e.g., child rearing and sex). However, it has often shared with other forms of critique a logic of explanation in terms of underlying and pregiven interests, here those of men and patriarchy. Hence it has operated in terms of an implicit division between the ways in which gender regulates women and the regulation of the characteristics of human males. The task for critical history, once again, is to reverse the lines of inquiry: to examine precisely how, and in what practices, gender was put together and distinguished. We need to locate the logics of explanation of pathology that problematized both the sexuality of men and the sexuality of women but in relation to different concerns. We need to examine not only the sorrows of the identification of women with domesticity and motherhood but the simultaneous construction of the pleasures and powers of the 'normal woman'. Women themselves, sometimes in alliance with men, sometimes to rescue and reform their fallen sisters, have been active participants here, often "heroes of their own lives" (Gordon, 1989). Within a critical history, the dividing practices organized around gender do not so easily assign females to a role as victims of history and males to role of orchestrators and beneficiaries of domination.

Recently, professional social historians and historians of science have begun to turn their attention to psychology, psychiatry, and psychoanalysis. These historians are frequently critical of the sociological accounts to which I have been referring, regarding them as simplistic, generalizing, ignorant of the fine detail of the historical record, and so forth. They have embarked on a lengthy project of historical rectification. No one can doubt the scholarship of the best of this work, and I have drawn extensively upon it. But I would suggest that it is in danger of missing the point. The questions that critique addressed through history were not themselves historical. First, there were sociological questions. These tried, in one way or another, to analyze the psychological sciences as bodies of belief, institutions, and techniques whose nature and emergence could be understood within a global social context. Second, there were political questions. These involved interrogating these scientific and technical exercises as systems of domination, and asking what forms of power they manifested and instantiated. Third, there were ethical questions. How were we to evaluate these new disciplines? This partly took the form of analyses of truth and falsity, of the scientific credentials of psychiatry and psychology, which were linked in these accounts to an interrogation of their humanity and effectiveness. It also took the form of a critical evaluation of the forms of life and systems of value to which the psychological disciplines had become attached.

The critiques frequently grounded their answers to such questions in analytic notions, which seem to me to be misguided. Nonetheless such sociological, political, and ethical questions are of abiding importance. A critical history of psychology and psychiatry and their allied technologies might best be undertaken through treating the very existence of these fields of knowledge and practice as a problem to be explained, and through locating their functioning in relation to a wider field of systems of social regulation, political domination, and ethical judgment. For psychology, like the other 'human' sciences, has played a fundamental role in the creation of the kind of present in which we in 'the West' have come to live. To address the relations between subjectivity, psychology, and society from this perspective is to examine those fields in which the conduct of the self and its powers have been linked to ethics and morality, to politics and administration, and to truth and knowledge. For such societies have been constituted, in part, through an array of plans and procedures for the shaping, regulation, administration of the self, which, over the past two centuries, has been inescapably bound to knowledges of the self. And psychology – indeed all the psy knowledges – have played a very significant role in the reorganization and expansion of these practices and techniques which have linked authority to subjectivity over the past century, especially in the liberal democratic polities of Europe, the

United States, and Australia. A lengthy historical research program is, in my view, neither necessary nor sufficient to address these concerns. So how might a critical history be undertaken?

The construction of the psychological

Up until quite recently, historical studies of psychology tended to operate in terms of some rather clear distinctions. There was a domain of 'reality' which psychology sought to know, but which existed independent of it – specified in various ways as the psyche, consciousness, human mental life, behavior, or whatever. There was the domain of 'psychology' – which again varied from account to account, but it generally consisted of psychologists or their precursors, theories, beliefs, books and articles, experiments, and the like. And there was the domain of 'society', construed either as 'culture' or 'world views' or processes such as 'industrialization' – which acted as a kind of backdrop to these attempts. Such histories sometimes asked questions about the relationships between psychology and society – how had 'social' phenomena such as religion, prejudice, or even institutional arrangements such as universities and professions, affected or influenced the development of psychology. And they sometimes asked how psychological theories and practitioners had affected society – how and where had they been 'applied', to what phenomena, and with what success. But they seldom, if ever, asked questions concerning the relations between the object of psychological knowledge – the mental life of the human individual, subjectivity – and psychological knowledge itself.

Recently a number of writers have successfully challenged these separations. Psychology, it has been shown, cannot be regarded as a given domain, separate from something called 'society' – the processes by which its truths are produced are constitutively 'social'. And, further, the object of psychology cannot be regarded as something given, independent, that preexists knowledge and which is merely 'discovered'. Psychology constitutes its object in the process of knowing it. One version of this line of argument has become known as 'social constructionism', and it has been deployed in a large number of studies of psychology (the classic statement is Gergen, 1985a; see also the essays collected in Gergen and Davis, 1985; Parker and Shotter, 1990; Burman, 1994; Morawski, 1988a). Social constructionist arguments, by and large, begin from a number of explicit or implicit propositions concerning knowledge. Knowledge is 'underdetermined' by experience, so that the world must be understood in terms that are the product of culture. Hence these understandings are dependent not on the nature of reality or the empirical validity of the propositions, but on social processes. These processes are social and historically variable and thus so is what counts as knowledge. Forms of knowledge are embodied in the linguistic interchanges and other interactions between individuals. Thus characteristics, capacities, processes, and the like habitually attributed to human beings in our culture or other cultures – childhood, love, the concept of the self, repertoires of emotions, femininity, motherhood, hostility, aggression – are more properly understood as the outcome of these social and interactional processes of construction.

These constructionist arguments in psychology have been developed in a number of directions that have a bearing on the history of the discipline. For some, they imply that the very object of psychology is itself historical. Psychology cannot attain universality in its laws for many reasons, no doubt, but ultimately because its very object – human psychology – changes as culture changes and is changed in part by psychology itself. This is not only because human dispositions are themselves historical and shaped by, among other things, the cultural take-up of psychological ideas in child-rearing practices and so forth. It is also because the languages that humans interpret themselves through, and thus construct themselves by, are subject to historical change and impacted upon by psychology itself (Gergen, 1985a, 1985b). For others, it is through historical investigation that one can examine the detailed and complex negotiations through which certain techniques of experimentation, forms of explanation, modes of argumentation have been accepted as defining the discipline of psychology, and through which the 'subject' of psychology is 'socially constructed' both in the sense of the construction of the domain of sanctioned knowledge and theory, and in the sense of the construction of its thought object, the human subject (Danziger, 1988, 1990; Morawski, 1988b). In a third approach, many have argued that what has been 'socially constructed' should be 'deconstructed'. Social construction here refers to a complex of "interpersonal, cultural, historical and political" processes – including psychology itself – that produce the objects that psychology studies, such as 'the child' or 'the mother' in relation to certain strategies of power or domination, and deconstruction to anything from a generic form of scrutiny and critique to a formal analytic method of disclosing the founding oppositions and absences upon which certain philosophies or forms of knowledge are grounded (Burman, 1994, p. 6; cf. Sampson, 1989; Parker and Shotter, 1990).

There is much to be learned from these studies. However, the *critical* purchase of these claims concerning the 'social construction' of psychology and its objects often rests upon an attack on implicit or explicit enemies – empiricism and positivism. That is to say, the rhetorical force of the argument that 'the child', 'the mother', 'the self', 'aggression', and the like are constructed depends on an antagonist who has claimed that they are 'discovered' – there, in reality, waiting to be uncovered and revealed by science. Of course, it is entirely understandable that psychologists critical of their discipline should pose their arguments in this form, given the way in which a certain image of science, of the logics of investigation, of experimentation, discovery, statistical validation, and so forth dominated psychological inquiry, especially in the American tradition over the middle decades of this century. But perhaps the repetition of the claim that 'x is not given in reality, but socially constructed' and the invocation of the imaginary enemy of positivism may now actually be an obstacle to critical inquiry. In scientific domains less racked by anxieties about their own status and respectability, philosophers and historians of science have long accepted that scientific truth is a matter of construction. What, then, if anything, distinguishes the 'constructions' in which psychology has participated from those which have been constitutive of other fields of scientific knowledge?

Phenomenotechnics

Let me begin by reflecting upon what is meant when it is argued that the object of knowledge is 'constructed'. Gaston Bachelard's writings on quantum physics, relativity, and non-Euclidean geometry can help us approach this question (Bachelard, [1934] 1984: all the quotations that follow come from pp. 12–13). Like Nietzsche, for Bachelard "everything crucial comes into being only 'in spite'. . . . Every new truth comes into being in spite of the evidence; every new experience is acquired in spite of immediate experience." And for Bachelard, this means that the activity of science engages in the 'construction' of new realms of scientific objectivity: science entails a break *away from the given,* from the world that experience appears to disclose to us.

In *The New Scientific Spirit,* Bachelard argues that scientific reason is necessarily a break away from the empirical. Science, he claims, should be understood not as a phenomenology but as 'phenomenotechnology': "It takes its instruction from construction." That is to say, science is not a mere reflection upon or rationalization of experience. Science, and Bachelard is being both descriptive and normative here, entails the attempt to produce in reality, through observation and experimentation, that which has *already* been produced in thought. In scientific thought, "meditation on the object always takes the form of a project. . . . Scientific observation is always polemical; it either confirms or denies a prior thesis, a pre-existing model, an observational protocol." And experimentation is essentially a process by which theories are materialized by technical means. For "once the step is taken from observation to experimentation, the polemical character of knowledge stands out even more sharply. Now phenomena must be selected, filtered, purified, shaped by instruments; indeed it may well be the instruments that produce the phenomena in the first place. And the instruments are nothing but theories materialized."

For Bachelard, then, reality should not be understood as some kind of primitive given: "every fruitful scientific revolution has forced a profound revision in the categories of the real" (ibid., 134). Indeed, Bachelard's notion of epistemological obstacles and his project for a 'psychoanalysis' of scientific reason arise from his injunction that science needs to exercise a constant vigilance against the *seduction of the empirical,* the lure of the given that serves as an impediment to the scientific imagination. This imperative opens up a key difference with the 'Anglo-American' analytics of 'constructionism'. Many contemporary Anglo-American constructionists seek to reveal the constructive character of scientific knowledge in order to 'deconstruct' it. They point to the ways in which scientific reality is produced by means of theory-laden instruments, techniques, and inscription devices in the course of an 'ironizing' or even a 'debunking' assault on the very idea of science. Much to the discomfort of those who propound these views, however, such a critique of science paradoxically saves empiricism: it bases itself on the very territory that it seeks to reproach. For its radical colors depend on the maintenance of an ideal of truth as that which *would be* grounded in the empirical. Only on this basis can it castigate all truth claims that are not so grounded: that are based on observations colored by theory and apparatus, on 'interpretation' dependent upon assumptions, on the attribution of 'mental processes' that go beyond the visible, hearable data of human interchanges. But within the more sober Bachelardian tradition, to point to the constructed nature of scientific objectivity is not to embarrass or debunk the project of science, not to 'ironize' it or 'deconstruct' it, but to *define* it. Contrary to all forms of empiricism – whether philosophically grounded or based upon a valorization of 'lay' knowledge and 'everyday experience' – for Bachelard, scientific reality is *not* in accordance with 'everyday thought': its objectivity is achieved and not merely 'experienced'. Contemporary scientific reality – and this goes for a science like psychology as much as any other – is the inescapable outcome of the categories we use to think it, the techniques and procedures we use to evidence it, the statistical tools and modes of proof we use to justify it.

From this perspective, to argue that the objects that appear within a particular domain of knowledge are constructed does not amount to a delegitimation of its scientific pretensions. It is merely the basis from which we become able to pose questions concerning the means of construction of these new domains of objectivity and their consequences. It is here that a second lesson can be drawn from Bachelard's arguments. Construction is not a matter of 'discourse' or language. It is a practical and technical matter (Hacking, 1990). It is this line of Bachelardian thought that has been extended by recent studies of science as technical, a matter of laboratories, apparatuses, inscriptions, charts, graphs, experiments, techniques, types of judgment, the circulation of knowledge though institutional devices of journals and conferences, the rhetorical and other procedures that stabilize facts and explanations (see especially Latour, 1988). The objects of a science, and psychology is no exception, are brought into existence by the assembling of these elements into a complex and heterogeneous network, many of the parts of which come from elsewhere and are stabilized by being locked into other circuits of activity, technique, and artifact. Those activities that we call science, and the objects of knowledge and systems of explanation and judgment they produce, are thus not merely matters of the elaboration of systems of meaning. Hence it is a vain exercise to seek to 'deconstruct' them by revealing the processes upon which their truth claims depend: something unsayable may lie at the heart of knowledge, but it is neither its origin nor its death knell.

One constructionist tendency in critical psychology has focused on the deployment of terms for psychological entities such as emotions, feelings, attitudes, and so forth in linguistic interchanges between human actors (e.g., Potter and Wetherell, 1984). Such approaches portray human individuals as agents, seeking to bring off their lives with the aid of the sense-making resources available to them, in particular the sense-making resources made available by language, though often, no doubt, unaware of how they do this and of the conventions and repertoires that constrain them. The psychological construction of reality is here examined though analysis of conversations of various types – between lay people, or between lay people and professionals, examining the sequencing, turn taking, membership categorization within these transcripts, seeking to identify how the parties mutually constructed a version of events, entailing certain kinds of explanation that posited a particular form of disturbed self, or a self with emotions or attitudes, lying behind the events, and then adduced this self as the explanation of them. Such analyses focus on the flexibility of the resources drawn on by the participants, the context-bound and indexical characteristic of much of the talk, the various forms in which persons were constructed by themselves or their interlocutors in order to attribute blame, to excuse, to accredit or discredit selves (cf. Burman and Parker, 1994). But the lines of investigation suggested here imply that there are conditions of making sense that go beyond the speaking subject and what is said. These provide the conditions under which it is possible for a person to take up the position of speaking subject, to identify oneself with the 'I' in one's speech, the set of relationships of sequencing, substitution, and association and difference that enable a particular sequence of sounds to have sense (Benveniste, 1971, esp. chap. 21; I extend this argument in Chapter 8). Discourses are not merely 'meaning systems', but are embodied within complex technical and practical associations and devices that provide 'places' that human beings must occupy if they are to have the status of subjects of particular sorts, and which immediately position them in certain relations with one another and with the world of which they speak (Foucault, 1972a).

Analyses conducted from this perspective proceed under radically different epistemological and methodological auspices to the Anglo-American tradition. First, there is a challenge to primacy of 'what is said' in relation to the conditions that make certain forms of utterance possible and intelligible: as Michel Foucault once put it in another context: "what does it matter who is speaking – someone has said . . ." (Foucault, 1969). There is also a challenge that one might term the 'metaphysics of presence', the epistemological doctrine that underpins Anglo-American constructionism and which leads to its fetishism of what is said – the hearable, that which appears to be immediately present to the consciousness or experience of subject and analyst alike – and its deprecation of explanation, which goes beyond 'the empirical evidence'. For what is present in the form of a sound, a statement, a sign has meaning and intelligibility only in relation to a set of discursive and technical connections that are absent, but which make that utterance possible. Hence there is a further challenge to the privilege accorded to the human subject in this business of construction: analysis must focus on the relations that provide the possibility of acting as a speaking subject of a particular type.

More positively, such analyses insist that psychology be understood not as a system of meaning or even as a 'discourse' but as technological. This term should be understood in the sense in which I have used it previously. Thus by a technology I mean an ensemble of arts and skills entailing the linking of thoughts, affects, forces, artifacts, and techniques that do not simply manufacture and manipulate, but which, more fundamentally, order being, frame it, produce it, make it thinkable as a certain mode of existence that must be addressed in a particular way. Psychology is technological in a number of senses. First, it seems to me useful to regard language itself – and hence the language of psychology – as constituting certain 'intellectual techniques', which renders reality thinkable in particular ways by ordering it, classifying it, segmenting it, establishing relations between elements, enabling it to become amenable to thought. Language – here psychological theories, concepts, entities, explanations – makes up a kind of intellectual machinery which can render the world amenable to being thought but only under certain descriptions. Further, psychology, like other disciplines, is not merely a complex of language but an array of techniques of inscription, procedures for entering aspects of the world into the sphere of the thinkable in the form of observations, graphs, figures, charts, diagrams, notations of various sorts (Lynch, 1985; Latour, 1986b; see my discussion in Chapter 5). These 'make up' the objects of psychological discourse by rendering them remarkable in particular ways. Third, psychology is intrinsically bound to 'human technologies'. It forms a part of the practical rationalities of assemblages that seek to act upon human beings to shape their conduct in particular directions – assemblages such as those of the legal apparatus, of schooling, of child rearing, even of spiritual guidance. The historical reality of psychological entities emerges, that is to say, neither from a prediscursive sphere of nature, nor from cultural mutations in patterns of meaning, but from the technical and practical organization of procedures for thinking, inscribing, and intervening upon human beings in heterogeneous assemblages of thought and action. How then might such a critical investigation of the practical, technical, and discursive construction of psychological entities proceed?

Regimes of truth

Whatever its insights into the technical and material character of the scientific activity, the Bachelardian model is too benign to account for the construction of psychological objectivity. Truth is not only the outcome of construction, but of contestation. There are battles over truth, in which evidence, results, arguments, laboratories, status and much else are deployed as resources in the attempts to win allies and force something into the true (for what follows, cf. Foucault, 1972a, 1972b, 1978; Latour, 1988). Truth, that is to say, is always enthroned by acts of violence. It entails a social process of exclusion in which arguments, evidence, theories, and beliefs are thrust to the margins, not allowed to enter 'the true'. If one wants an example of this, one need look no further than the 'truth battles' that have characterized the relation of psychology and psychoanalysis in different national territories: battles over the status of its theories, its results, its discoveries, its practitioners. These battles over truth are not abstract, for truth inheres in material forms. To be in the true, facts and arguments must be permitted to enter into complex apparatuses of truth – scholarly journals, conferences, and the like – which impose their own norms and standards upon the rhetorics of truths. Truth entails an exercise in alliances and persuasion both within and without the bounds of any disciplinary regime, in which process an audience for truth can be identified and enrolled. And truth entails the existence of a mode of human existence within which such truth might be feasible and operative.

From such a perspective, we can explore the particular conditions under which psychological arguments have been allowed 'in the true'. The notion of 'translation', developed in the work of Bruno Latour and Michel Callon, is helpful in understanding these processes: "By translation we understand all the negotiations, intrigues, calculations, acts of persuasion and violence, thanks to which an actor or force takes, or causes to be conferred on itself, authority to speak or act on behalf of another actor or force; 'Our interests are the same', 'do what I want', 'you cannot succeed without me'" (Callon and Latour, 1981, p. 279). It is through such processes of translation, Callon and Latour suggest, that very diverse entities and agents – laboratory researchers, academics, practitioners, and social authorities – come to be linked together (Callon, 1986; Latour, 1986b). Actors in locales separated in time and space are enrolled into a network to the extent that they come to understand their situation according to a certain language and logic, to construe their goals and their fate as in some way inextricable.

To understand the 'construction of the psychological' does indeed entail an investigation of the ways in which networks were formed that operated under a certain 'psychological' regime of truth. However, I think that Callon and Latour oversimplify, for they imply that networks are always established on the basis of a 'will to power' on the part of individual or collective actors, and that they entail an exercise of 'domination' effected by particular centers (cf. Latour, 1984). But these 'truth battles' are not 'zero sum games' where what is won by one side is lost by another. Rather, through a complex of seductions, associations, problematizations, and machinations, certain forms of thinking and acting propagate because they appear to be solutions to the problems and decisions confronting actors in a variety of settings (cf. Miller and Rose, 1994). Nonetheless Callon and Latour are right in rejecting accounts of such processes posed in terms of either the insipid notion of 'diffusion of ideas' or the cynical notion of satisfaction of 'social interests'. Kurt Danziger's meticulous examination of the relation between the deployment of psychology in these practical domains and the psychology of the laboratory has illustrated clearly some of the political and rhetorical processes through which such alliances have been formed, and their consequences for what is to count as valid psychological knowledge (Danziger, 1990). A political and rhetorical labor is involved in constructing a 'translatability' between the laboratory, the textbook, the manual, the academic course, the professional association, the courtroom, the factory, the family, the battalion, and so on – the diverse loci for the elaboration, utilization, and justification of psychological statements (cf. the essays collected in Morawski, 1988a).

In the case of psychology, we can distinguish a number of different tactics through which translation has occurred, simultaneously stabilizing psychological thought and creating a psychological territory. First of all, this has entailed persuasion, negotiation, and bargaining between both social and conceptual authorities, with all the calculations and trade-offs one might expect. Second, it has involved fashioning a mode of perception in which certain events and entities come to be visualized according to particular images or patterns. Third, it has been characterized by the circulation of a language in which problems come to be articulated in certain terms, accounted for according to certain rhetorics, objectives, and goals identified according to a certain vocabulary and grammar. Fourth, the enrolling of agents into a 'psychologized' network entails establishing problem-solution linkages: connections between the nature, character, and causes of problems facing different individuals and groups – doctors and teachers, industrialists and politicians – and certain things that can count as actual or potential solutions to these problems.

Consider, for example, the growth of the language and strategies of intelligence in the early years of the twentieth century, or of mental hygiene in the 1920s and 1930s (both discussed in Rose, 1985a). In each case, what one observes is the establishment of mobile and thixotropic associations between various agents – academic psychologists, professionals such as teachers and doctors, politicians and political pressure groups and organizations, industrialists, well-meaning individuals – in which they seek to enhance their own capacity for action and persuasion by 'translating' the resources provided by the association so that these may work to their own advantage. Through adopting shared problem definitions and vocabularies of explanation, loose and flexible linkages can be put in place between those who are separated spatially and temporally, and between events in spheres that remain formally distinct and autonomous. These alliances between the researchers and the practitioners, the producers and the consumers of psychological knowledge, so essential to its construction, impart a particular character to the process of construction of what will count as psychological knowledge.

Disciplinization

The 'disciplinization' of psychology from the mid-nineteenth century onward was inextricably linked to the possibility of the building of such alliances. What one sees in the disciplinization of psychology, however, is really rather specific: the conditions for such a disciplinary stabilization lay in the elaboration of a whole range of techniques and practices for the discipline, surveillance, and formation of populations and the human beings that make them up (Gordon, 1980, p. 239). These alliances made a positive knowledge of 'man' possible. 'Man' became, as it were, an imaginary point of reference, the universe within which all the classifications and categorizations of age, race, sex, intelligence, character, pathology were drawn. The conditions of the birth of such a positive knowledge shaped it in certain very significant respects, discussed in other chapters of this book. Here I wish to focus upon some different issues.

First, perhaps, we can identify the ways in which certain norms and values of a technical nature came to define the topography of psychological truth. Most significant here were *statistics* and *the experiment.* The constitutive role of 'tools' and 'methods' in the establishment of a psychological regime of truth requires us to redraw Bachelard's diagram of the relation between thought and technique. In the construction of psychological truth, the technical means available for the materialization of theory have played a determining and not a subordinate role. The technical and instrumental forms that psychology has adopted for the demonstration and justification of theoretical propositions have come to delimit and shape the space of psychological thought itself. The disciplinary project of psychology, over the fifty years that followed the establishment of the first psychological laboratories, journals, and societies in the late nineteenth century, was accomplished, to a large extent, in a process that required psychology to jettison its previous modes of justification and adopt 'truth techniques' already established in other domains of positive knowledge.

The two truth techniques that were preeminent here were 'statistics' and 'the experiment' (Rose, 1985a, chap. 5; Danziger, 1990; Gigerenzer, 1991). Both exemplify not merely the alliances formed by psychology with other scientific disciplines, but also the reciprocal interplay between the theoretical and the technical. Statistics, of course, emerged originally as 'science of state', the attempt to gather numerical information on events and happenings in a realm in order to know and govern them – the formation of a lasting relation between knowledge and government. Ian Hacking has argued convincingly that, in the course of the nineteenth century, the earlier assumption that statistical laws were merely the expression of underlying deterministic events gave way to the view that statistical laws – the laws of large numbers formulated in the 1830s and 1840s by Poisson and Quetelet – were laws in their own right that could be extended to natural phenomena (Hacking, 1990). A conceptual rationale was thus constructed for the claim that regularity underlay the apparently disorderly variability of phenomena.

In the first thirty years or so of psychology's disciplinary project, from the 1870s to the early years of this century, programs for the stabilization of psychological truths went hand in hand with the construction of the technical devices necessary to demonstrate that truth. In the work of Francis Galton, Karl Pearson, Charles Spearman, and others, and from the notion of a 'normal distribution' to the devices for calculating correlations, the relation between the theoretical and the statistical was an internal one. Statistics were instruments that both materialized the theory and produced the phenomena that the theory was to explain. Statistical techniques began as a *condensation of the empirical* and were then reshaped in such a way that they became a *materialization of the theoretical.* Within a remarkably short period of time, however, they became detached from the specific conceptual rationales that underpinned them: by the 1920s, statistical laws appeared to have an autonomous existence, which was merely accessed by statistical devices. Statistical tests appeared as an essentially neutral means for the demonstration of truth deriving from a universe of numerical phenomena, which, because untainted by social and human affairs, could be utilized to adjudicate between different accounts of such affairs. Not only psychology, but also the other 'social sciences' would seek to utilize such devices to establish their truthfulness and scientificity, to force themselves into the canon of truths, to convince sometimes skeptical audiences of politicians, practitioners, and academics of their veridicality, to arm those who professed them with defenses against criticisms that they were merely dressing up prejudice and speculation in the clothes of science. From this point forward, the means of justification come to shape that which can be justified in certain fundamental ways: statistical norms and values become incorporated within the very texture of conceptions of psychological reality (cf. Gigerenzer, 1991).

'The experiment' was also to be embraced by psychology as a means of disciplining itself, of lashing together the various constituencies of practitioners, journal editors, funding bodies, fellow academics, and university administrators into the alliances necessary to force itself into the apparatus of truth. The interminable debate over the relations between the psychological 'sciences' and the 'natural sciences' is better understood if it is removed from the realm of philosophy and relocated in a question of technique (this argument is based on Danziger, 1990). In seeking to establish their credibility with necessary but skeptical allies, British and American psychologists in the early decades of this century abandoned their attempts to craft an investigative method that answered to a conception of the human subject of investigation as an active participant in the process of generation and validation of psychological facts. The 'experimental method' in psychology was not merely sanctified through the attempt to simulate a model for the production and evaluation of evidence derived from (naive) images of laboratories in physics and chemistry. It also arose out of a series of practical measures for the generation and stabilization of data in calculable, repeatable, stable forms. These included the establishing of psychological laboratories as the ideal site for the production, intensification, and manipulation of psychological phenomena; the separation of the experimenter endowed with technical skills and the subject whose role was merely to provide a source of data; the attempt to generate evidence in the form of inscriptions amenable to comparison and calculation; and the like. As one particular form of the psychological experiment became institutionalized and policed by the emerging disciplinary apparatus, the social characteristics of the experimental situation were naturalized. The norms of the experimental program had, as it were, merged with the psychological subject itself; in the process the object of psychology was itself disciplined. It became 'docile'; it internalized the technical means to know it in the very form in which it could be thought (Rose, 1990, chap. 12; cf. Lynch, 1985). Psychological truths here were no simple materialization of theory; indeed the reverse is probably closer to the truth. The disciplinization of psychology as a positive science entailed the incorporation of the technical forms of positivity into the object of psychology – the psychological subject – itself.

Psychologization

The 'disciplinization' of psychology was intrinsically bound to the 'psychologization' of a range of diverse sites and practices, in which psychology came to infuse and even to dominate other ways of forming, organizing, disseminating, and implementing truths about persons. The regulatory and administrative requirements of an actual or potential constituency of social authorities and practitioners played a key role in establishing the kinds of problems that psychological truths claim to solve and the kinds of possibilities that psychological truths claim to open. No single process was involved here; to write the genealogy of contemporary psychology one would need to examine in detail the diverse sites that were psychologized – factories, courtrooms, prisons, schoolrooms, bedrooms, colonial administration, urban spaces – and the different images and technologies of human subjects that were established and deployed within them (I make a start at this myself in Rose, 1990). For psychologization does not imply that a single model of the person was imposed or adopted in a totalitarian manner, indeed psychology's celebrated 'nonparadigmatic' character ensures a kind of perpetual contestation over the characteristics of personhood. Consider, for example, the differences between the nineteenth-century psychological characterization of gender in the schoolroom, of race in relation to the inheritance of intelligence, of criminality in the criminal courts acting upon adults and children, of reputation in relation to the legal treatment of libel and slander, and so forth. This variability in psychological ways of 'making up' persons is a key to the wide-ranging power of psychology, for it enables the discipline to tie together diverse sites, problems, and concerns. The social reality of psychology is not as a kind of disembodied yet coherent 'paradigm', but as a complex and heterogeneous network of agents, sites, practices, and techniques for the production, dissemination, legitimation, and utilization of psychological truths.

The production of psychological 'truth effects' is thus intrinsically tied to the process though which a range of domains, sites, problems, practices, and activities have 'become psychological'. They 'become psychological' in that they are *problematized* – that is to say, rendered simultaneously troubling and intelligible – in terms that are infused by psychology. To educate a child, to reform a delinquent, to cure a hysteric, to raise a baby, to administer an army, to run a factory – it is not so much that these activities entail the utilization of psychological theories and techniques than that there is a constitutive relation between the character of what will count as an adequate psychological theory or argument and the processes by which a kind of psychological visibility may be accorded to these domains. The conduct of persons becomes remarkable and intelligible when, as it were, displayed upon a psychological screen, reality becomes ordered according to a psychological taxonomy, and abilities, personalities, attitudes, and the like become central to the deliberations and calculations of social authorities and psychological theorists alike.

Institutional epistemology

Michel Foucault remarks somewhere that the psy knowledges have a "low epistemological profile." The boundaries between that which psy organizes in the form of positive knowledge and a wider universe of images, explanations, meanings, and beliefs about persons are indeed more 'permeable' in the case of psy than, say, in the field of atomic physics or molecular biology. But we should not merely pose this question of permeability in the terms of the history of ideas, where scientific discourses are seen to partake of metaphors or key notions that are widely socially distributed. I would prefer to examine this relation at a more modest and technical level. In the case of psy knowledges, that is to say, there is an interpenetration of practicability and epistemology. We have already examined some of these relations, but we can investigate the 'practical' constitution of psychological epistemology in another way. Bachelard argues that scientific thought does not work on the world as it finds it; the production of truth is an active process of intervention into the world. But there is something characteristic about the conditions under which psychological truths have been produced. Psychological epistemology is, in many respects, an *institutional epistemology* (cf. Gordon, 1980): the rules governing what can count as knowledge are themselves structured by the institutional relations in which they have taken shape.

Michel Foucault utilized the notion of *surfaces of emergence* to investigate the apparatuses within which the troubles or problem spaces condensed that were later to be rationalized, codified, and theorized in terms such as disease, alienation, dementia, neurosis (1972a). Such apparatuses – for example, the family, the work situation, the religious community – have certain characteristics. They are normative, and hence sensitive to deviation. They provide the focus for the activity of authorities – such as the medical profession – who will scrutinize and adjudicate events within them. And they are the locus for the application of certain grids of specification for dividing, classifying, grouping, and regrouping the phenomena that appear within them.

As far as psychology was concerned, it was within the prison, the courtroom, the factory, the schoolroom – institutional spaces which collected people together and judged them in terms of organizational requirements such as timekeeping and obedience – that the objects were formed that psychology would seek to render intelligible (Foucault, 1977; Rose, 1985a; cf. Smith, 1992). Psychology disciplined itself through the codification of the vicissitudes of individual conduct as they appeared within the apparatuses of regulation, administration, punishment, and cure as they took their modern shape in the second half of the nineteenth century and the first decades of the twentieth. Within these apparatuses, psychology would align itself with institutional *systems of visibility*. That is to say it was the normativity of the apparatus itself – the norms and standards of the institution, their limits and thresholds of tolerance, their rules and their systems of judgment – that conferred visibility upon certain features and illuminated the topography of the domains that psychology would seek to render intelligible. Its conceptions of intelligence, personality, attitudes, and the like would establish themselves as truthful only to the extent that they could be simultaneously practicable, translated back into the disciplinary requirements of the apparatus and its authorities. Hence, to return to Bachelard, the psychologist's meditation on his or her scientific object has not taken the form of a polemical intervention into reality to realize a scientific thesis. Rather, it has been characterized by a range of attempts to rationalize an already existing domain of experience and render it comprehensible and calculable (cf. this volume, Chapter 4).

However, rendering a preexisting problem space comprehensible and calculable in psychological terms does not leave it in its original state. Psychological ways of seeing, thinking, calculating, and acting have a particular potency because of the *transformations* that they effect on such problem spaces. They confer a certain *simplification* upon the range of activities that authorities engage in when they deal with the conduct of conduct. If one considers, say, the transformation of 'social work' in the 1950s and 1960s or the rise of 'person-centered' approaches in general medical practice in the 1960s and 1970s, one can see how psychology, in 'rationalizing' the practice of other 'specialists', simplifies their diverse tasks by rendering them as all concerned with different aspects of the personhood of the client or patient. Psychology not only offers these authorities a plethora of new devices and techniques – for the allocation of persons to tasks, for the arrangement of the minutiae of the technical arrangements of an institution, for architecture, timetabling, and spatial organization, for the organization of working groups, leadership, and hierarchy. It also accords these mundane and heterogeneous activities a coherence and a rationale, locating them within a single field of explanation and deliberation: they are no longer ad hoc, but purport to be grounded in a positive knowledge of the person. And, in the process, the very notion of authority, and of the power invested in the one who exercises it, is transformed.

The power of psychology thus initially derived from its capacity to organize, simplify, and rationalize domains of human individuality and difference that emerged in the course of institutional projects of cure, reform, punishment, management, pedagogy, and the like. But, in simplifying them, it transforms them in certain fundamental ways.

The techne *of psychology*

Suppose we consider psychology not as merely a body of thought but as a certain form of life, a mode of practicing or acting upon the world. We could then seek to identify what one might term the *techne* of psychology: its distinctive characteristics as skill, art, practice, and set of devices. I discuss this proposition in more detail elsewhere (see Chapter 4); here I would like to highlight just three aspects of this *techne,* three dimensions of the relations between psychology, power, and subjectivity: first, a transformation in rationales and programs of *government;* second, a transformation in the legitimacy of *authority;* and, third, a transformation in *ethics.*

Government

By government I refer not to a particular set of political institutions, but to a certain mode of thinking about political power and seeking to exercise it: the territory traced out by the multitude of schemes, dreams, calculations, and strategies for 'the conduct of conduct' that have proliferated over the last two centuries (Foucault, 1991). Over the course of the twentieth century, psychological norms, values, images, and techniques have increasingly come to shape the ways in which various social authorities think of persons, their vices and virtues, their states of health and illness, their normalities and pathologies. Objectives construed in psychological terms – normality, adjustment, fulfillment – have been incorporated into programs, dreams, and schemes for the regulation of human conduct. From the 'macro', the apparatuses of welfare, security, and labor regulation, to the 'micro', the individual workplace, family, school, army, courtroom, prison, or hospital, the administration of persons has taken a psychological coloration. Psychology has been embodied in the techniques and devices invented for the government of conduct and deployed not only by psychologists themselves but also by doctors, priests, philanthropists, architects, teachers. Increasingly, that is to say, the strategies, programs, techniques and devices, and reflections on the administration of conduct that Michel Foucault terms governmentality or simply government have become 'psychologized'. The exercise of modern forms of political power has become intrinsically linked to a knowledge of human subjectivity.

Authority

Psychology has been bound up with a transformation in the nature of social authority that is of fundamental importance for the kinds of society we live in in 'the West'. First, of course, psychology has itself produced a range of *new social authorities* whose field of operation is the conduct of conduct, the management of subjectivity. These new authorities, such as clinical, educational, and industrial psychologists, psychotherapists, and counselors, claim social powers and status on account of their possession of psychological truths and their mastery of psychological techniques. Second, and perhaps more significant, psychology has been bound up with the constitution of a range of *new objects and problems* over which social authority can legitimately be exercised, and this legitimacy is grounded in beliefs about knowledge, objectivity, and scientificity. Notable here are the emergence of *normality* as itself the product of management under the tutelage of experts, and the emergence of *risk* as danger *in potentia* to be diagnosed by experts and managed prophylactically in the name of social security (cf. Castel, 1991).

Third, the infusion of psychology into already existing systems of authority – that of the commander in the army, the teacher in the school, the manager in the factory, the nurse in the psychiatric hospital, the magistrate in the courtroom, the prison officer in the jail – has transformed them. These forms of authority accumulate a kind of *ethical basis,* through their infusion with the terminology and techniques attributable (in however a dubious and disingenuous manner) to psychology. Authority, that is to say, becomes ethical to the extent that it is exercised in the light of a knowledge of those who are its subjects. And the nature of the exercise of authority is simultaneously transformed. It becomes not so much a matter of ordering, controlling, commanding obedience and loyalty, but of improving the capacity of individuals to exercise authority over themselves – improving the capacity of schoolchildren, employees, prisoners, or the soldiers to understand their own actions and to regulate their own conduct. The exercise of authority, here, becomes a therapeutic matter: the most powerful way of acting upon the actions of others is to change the ways in which they will govern themselves.

Ethics

The history, sociology, and anthropology of subjectivity has been examined in many different ways. Some authors, notably Norbert Elias, have tried to relate changing political and social arrangements, changing codes of personal conduct, and changes in the actual internal psychological organization of subjects (Elias, 1978). Others have sought to avoid any imputations of internal life to humans, treating linguistic and representational practices as simply repertoires of *accounts* that provide the resources through which subjects make sense, of the actions of themselves and others (Harré, 1983). I approach this issue from a rather different perspective: changing discourses, techniques, and norms that have sought to act upon the minutiae of human conduct, human comportment, and human subjectivity – not just manners but also desires and values – are located within the domain of *ethics.*

An examination of the *techne* of psychology along this ethical dimension does not address itself to 'morality' in the Durkheimian sense of a realm of values and its associated mode of producing social integration and solidarity. Rather, it investigates the ways in which psychology has become bound up with the practices and criteria for 'the conduct of conduct' (Foucault, 1988). For many centuries manuals concerning manners, books of advice and guidance, pedagogic and reformatory practices have sought to educate, shape, and channel the emotional and instinctual economy of humans by inculcating a certain ethical awareness into them. But over the past fifty years, the languages, techniques, and personnel of psychology have infused and transformed the ways in which humans have been urged and incited to become ethical beings, beings who define and regulate themselves according to a moral code, establish precepts for conducting and judging their lives, and reject or accept certain moral goals for themselves.

From this perspective, psychology's relation to the self should not be construed in terms of an opposition between etiolated psychological conceptions of the person and real, concrete, creative personhood. This was the theme of so many critiques of the psychology of intelligence, personality, and adaptation in the 1960s and is still a theme in the new 'humanist' psychologies. But it is more instructive to examine the ways in which psychology has participated in the construction of diverse repertoires for speaking about, evaluating, and acting upon persons that have their salience in different sites and in relation to different problems, and have a particular relationship to the types of self that are presupposed in contemporary practices for the administration of individuals (Rose, 1992a, reprinted in a revised version as Chapter 7 of this volume).

On the one hand, the person has been opened up, in diverse ways, to interventions conducted in the name of subjectivity: the calculable subject, equipped with relatively stable, definable, quantifiable, linear, normally distributed characteristics – the domains of intelligence, personality, aptitude, and the like; the motivated subject, equipped with an internal dynamic orientation to the world, with needs to be shaped and satisfied; the social subject, seeking solidarity, security, and a sense of worth; the cognitive subject in search of meaning, steered through the world by beliefs and attitudes; the psychodynamic subject, driven by unconscious forces and conflicts; the creative subject, striving for autonomy through fulfillment and choice, according meaning to its existence through the exercise of its freedom. In liberal democratic societies, norms and conception of subjectivity are pluralistic. But the condition of possibility for each version of the contemporary subject is the birth of the person as a psychological self, the opening of a space of objectivity located in an internal 'moral' order, between physiology and conduct, an interior zone with its own laws and processes that is a possible domain for a positive knowledge and a rational technique.

On the other hand, diverse fragments and components of psy have been incorporated into the 'ethical' repertoire of individuals, into the languages that individuals use to speak of themselves and their own conduct, to judge and evaluate their existence, to give their lives meaning, and to act upon themselves. This transforms what I term, following Foucault, our 'relation with ourselves' – the way in which we make our being and our existence intelligible and practicable, our ways of thinking about and enacting our passions and our aspirations, and identifying, coding, and responding to our disaffections and our limits.

The construction of the psychological

From this perspective, psychology is significant less for what it is than for what it does. Psychology, that is to say, has altered the way in which it is possible to think about people, the laws and values that govern the actions and conduct of others, and indeed of ourselves. What is more, it has endowed some ways of thinking about people with extra credibility on account of their apparent grounding in positive knowledge. In making the human subject thinkable according to diverse logics and formulas, and in establishing the possibility of evaluating ways of thinking about people by scientific means, psychology also makes human beings amenable to having certain things done to them by others. It also makes it possible for them to do new things to themselves. It opens people up to a range of calculated interventions, whose ends are formulated in terms of the psychological dispositions and qualities which determine how human individuals conduct themselves, interventions whose means are inescapably adjusted in the light of psychological knowledge about the nature of humans.

The aim of a critical history of psychology would be to make visible the profoundly ambiguous relations between the ethics of subjectivity, the truths of psychology, and the exercise of power. Such a critical history would open a space in which we could rethink the constitutive links between psychology – as a form of knowledge, a type of expertise, and a ground of ethics – and the dilemmas in the government of subjectivity that confront liberal democracies today.

3

Psychology as a social science

To many of its critics, psychology is an 'antisocial' science, focusing on the properties of individuals abstracted from social relations, reducing social issues to interpersonal ones, servicing an unequal society. There is much of value in such criticisms. But suppose we were to reverse our perspective, to view psychology as a profoundly *social* science. The subdiscipline of 'social psychology' would be located within a broader web of relations that connect even the most 'individualistic' aspects of psychology into a social field. It is not only that truth is a constitutively social phenomenon: like any other body of knowledge staking its claims in the commonwealth of science, the truths of psychology become such only as the outcome of a complex process of construction and persuasion undertaken within a social arena. It is also that the birth of psychology as a distinct discipline, its vocation and destiny, is inextricably bound to the emergence of the 'social' as a territory of our thought and our reality.

Governing social life

No doubt all humans are social animals. But the social territory is a historical achievement, a shifting and uncertain terrain that began to consolidate in Western societies in the nineteenth century (cf. Deleuze, in Donzelot, 1979). It is the terrain implied by such terms as social security, social welfare, social workers, and social services. The social is a matrix of deliberation and action, the object of certain types of knowledge, the location of certain types of predicaments, the realm traced out by certain types of apparatus, and the target of certain types of program and ambition. Psychology as a discipline – a heterogeneous assemblage of problems, methods, approaches, and objects – was born in this social domain in the nineteenth century, and its subsequent vicissitudes are inseparable from it. And psychology, as a way of knowing, speaking, calculating, has played a constitutive part in the formation of the social. As the human soul became the object of a positive science, human subjectivity and intersubjectivity became possible targets of government.

Government, in the sense in which I use it, is not a matter of the minutiae of political intrigue or the complex relations between politicians, civil servants, bureaucrats, pressure groups, and so on. But nor should it be understood in terms of 'the state', an omnipotent and omniscient entity extending its control from the center throughout the social body. I use the term here to refer to what Foucault termed 'governmentalities': "the ensemble formed by the institutions, procedures, analyses and reflections, the calculations and tactics that allow the exercise of this very specific albeit complex form of power, which has as its target population" (Foucault, 1979a, p. 20).

Governmentalities are combinations of *political rationalities* and *human technologies.* They are ways of construing the proper ends and means of political authority: the objects to which rule should be addressed, the scope of political authority, the legitimate methods it may use. And they are ways of seeking to operationalize such ambitions, devising techniques and constructing devices to act upon the lives and conduct of subjects, to shape them in desired ways (Foucault, 1979a, 1979b, 1981; Miller and Rose, 1990; Rose and Miller, 1992).

This general field of governmentalities began to take shape in eighteenth-century Europe. Previously the tasks of princes and rulers had largely been limited to the maintenance and augmentation of the power of the state through accumulation of wealth, raising of armies, and the exercise of sovereignty by the promulgation of laws and decrees. Over this period, however, it began to be argued that population, rather than simply territory, should be the object par excellence of political rule. The strength of the state came to be associated with the good order and correct disposition of the persons, goods, and forces within its territory. And the exercise of political power came to be seen as depending on procedures of rational calculation and planning, upon those who could develop methods to act upon individuals and populations, not just to avert evil, but also to promote good.

This distinctive combination was termed, in the eighteenth century, the 'science of police' (Schumpeter, 1954; Pasquino, 1978; Oestreich, 1982). The three dimensions of police mark out the space within which the discipline of psychology was to be born. There were the objectives of police, which concerned not just the minimization of lawbreaking and other harms, but simultaneous augmentation of the coffers of the state and the wealth of the population, the maximization of public tranquillity and the qualities of individuals. This went hand in hand with the elaboration of procedures for the collection of information on the realm to be governed, on all the capacities and resources that the population of a territory comprised. And this was linked to the invention of techniques for the administration of the population, in the form of regulations governing the good order, education, habits, and security of persons in various towns and regions.

The three dimensions of police constitute a diagram of governmentality as it would take shape over the next two centuries. Police entails ways of thinking about the population, ways of rendering it the object of political discourse and political calculation. It requires ways of knowing the population, instituting a vast enterprise of inquiry into its state and condition. And it demands the bringing into being of the mechanisms that can enable those in authority to act upon the lives and conducts of subjects. However, the totalizing and omniscient aspirations of the doctrines of police, significant as they were, were to be challenged by two correlative developments at the end of the eighteenth century. In the emergence of conceptions of the economy in the writings of political economists such as Adam Smith, and notions of the population as subject to its own laws and regularities in the writings of those like Thomas Malthus, a new domain emerges into thought. As Foucault puts it (1986d, p. 242):

> What was discovered at that time – and this was one of the great discoveries of political thought at the end of the eighteenth century – was the idea of *society*. That is to say, that government not only has to deal with a territory, with a domain, and with its subjects, but that it also has to deal with a complex and independent reality that has its own laws and mechanisms of disturbance. This new reality is society. From the moment that one is to manipulate a society, one cannot consider it completely penetrable by police. . . . It becomes necessary to reflect upon it, upon its specific characteristics, its constants and its variables.

Psychology would emerge, in its modern and disciplined form, in the course of such reflections. It would be *society* that would have to be known, would have to be administered, whose well-being was to be assured by calculated action, not as a space to be despotically mastered, but as a domain to be governed in relation to its own laws, processes, and characteristics. It is here that the second significant shift in political thought is pertinent – the emergence of 'liberalism'. Liberalism, in the sense in which I use the term, should not be understood as simply a school of political philosophy, but as a series of reflections on government that stressed its limits (cf. Gordon, 1991; Burchell, 1991). Government, in these forms of thought, must limit itself in relation to the need to respect the autonomy of 'civil society', 'the family', 'the market', 'the free citizen'. Yet the well-being of a nation depended on good public order, proper family regimes, honest and thriving trade in markets, responsible citizens. To govern in a liberal way was to try to reconcile these two principles: the dangers of governing too much with the dangers of not governing enough (Miller and Rose, 1990; Rose and Miller, 1992). Programs for the liberal government of society opened a space in which the psychological sciences would come to play a key role. For these sciences are intrinsically tied to strategies that, in their wish to govern subjects as responsible but free citizens, have found that they need to know them.

The knowledge of government

Government depends upon knowledge. Not simply the knowledge of statecraft, which had been the subject of innumerable books of advice to princes in classical antiquity and in the Middle Ages. But a positive knowledge of the domain to be governed, a way of rendering it into thought, so that it can be analyzed, evaluated, its ills diagnosed and remedies prescribed. Such 'representation' has two significant aspects: the articulation of languages to *describe* the object of government and the invention of devices to *inscribe* it. On each of these dimensions, psychology will play a key role.

The languages of government do not merely mystify domination or legitimate power: they make new sectors of reality thinkable and practicable. Only through language can the ends of government be formulated, by portraying their object as an intelligible field with identifiable limits within which certain characteristics are linked in a systematic manner. Whether it be a question of governing an economy, a society, an enterprise, a family, or oneself, the domains in question are realized, brought into existence, through the languages that 're-present' them, and the calculations, techniques, and apparatuses that these languages make conceivable (Braudel, 1985; Tribe, 1976; Forquet, 1980; Miller and O'Leary, 1987; Rose, 1988; Miller and Rose, 1990).

The vocabularies of the psychological sciences have made two distinct but related contributions to the exercise of government over the past century. First, they provided the terms that enabled human subjectivity to be translated into the new languages of government of schools, prisons, factories, the labor market, and the economy. Second, they constituted subjectivity and intersubjectivity as themselves possible objects for rational management, in providing the languages for speaking of intelligence, development, mental hygiene, adjustment and maladjustment, family relations, group dynamics, and the like. They made it possible to think of achieving desired objectives – contentment, productivity, sanity, intellectual ability – through the systematic government of the psychological domain (Rose, 1990).

The succession of vocabularies in which psychology has been articulated over this period has been affiliated to a sequence of *problematizations.* The schoolroom, the slum, the court, the army, the factory, the family each constituted *surfaces of emergence* upon which problems would take shape – racial degeneration, intellectual decline, juvenile delinquency, shell shock, industrial inefficiency, childhood maladjustment – that psychology would make its own (cf. Foucault, 1972a). In and around these sites, psychology would find its subjects, scrutinize and study them, seek to reform or cure them, and, in the process, elaborate theories of mental pathology and norms of behavior and thought. A psychological knowledge of normality did not precede these concerns with abnormality – quite the reverse. It was in terms of problems of pathology, identifying them and managing them, that psychology would begin to elaborate its theories of psychological normality (Rose, 1985a). And psychology, in turn, fed back into social and political thought, the diversity of its languages making it a fertile resource for transforming, in many different ways, the problematizations of human existence, and thus opening up new domains for thought and action.

In the nineteenth century, government addressed itself to a moral order articulated in a language drawn from associationist epistemology, moral philosophy, and the new psychological theories of mental pathology. This vocabulary construed individual conduct as shaped by the cumulative effects of exposure to virtuous or vicious influences in early life, these having a lasting effect on constitution and character and even communicable to future generations. For example, the juvenile delinquent had contracted bad habits from the corrupt moral milieu at the heart of the large towns. Such delinquency was preventable through the remoralization of the domestic environment, which thus became the focus of innumerable philanthropic projects in London, Manchester, and other British and North American industrial cities, ranging from campaigns to get the poor to marry to programs for the reform of domestic architecture. Hence, too, it was argued that the potential or actual delinquent could be reformed if removed from the corrupt influences of the home and its surroundings, and placed in a regulated and moral environment where old habits could be broken down and new ones constructed. The techniques invented for the 'moral treatment' of the insane could be reused in a new moral pedagogy for pathological children. And this language, and the ways of thinking and acting that it brought into being, extended beyond the reform of the pathological to the development of techniques for the moralization of children in home and school. Psychological vocabularies were coming to play a crucial role in the rationales and techniques of social government.

In the last decades of the nineteenth century, moral judgments as to the vicious, corrupt, and socially dangerous nature of criminality, insanity, delinquency, prostitution, and debauchery were provided with a medicopsychiatric rationale in the language of 'degeneracy' (Foucault, 1979b; Pick, 1989). No longer the remediable consequences of immersion in a corrupt moral milieu, all these social dangers were the varied expressions of an inherited degenerate constitution. Degeneracy was inherently cumulative, for the unsavory and immoral life favored by those with tainted constitutions would result in an even greater hereditary weakness being passed on to their offspring, a tendency exacerbated by the tendency of the degenerate to interbreed, and to produce many offspring. National well-being was under threat from such profligate breeding of those with degenerate constitutions, handing their degeneracy down from generation to generation. New positive knowledges of the pathological – criminal anthropology, individual psychology, psychiatry – construed the moral state of the population in these terms and were articulated within eugenic programs to chart it and curtail it, by detecting degenerates and limiting their breeding.

In the 1920s and 1930s, it was the more benign vocabulary of mental hygiene that defined the terms for the government of subjectivity (Rose, 1986a; Miller and Rose, 1988). The language of mental hygiene reconstrued phenomena from crime to industrial accidents as symptoms of mental disturbances having their origin in minor troubles of childhood, themselves arising from disturbances in the emotional economy of the family. Hence the minor problems of childhood – lying, quarreling, bed-wetting, temper tantrums – became significant, not in themselves, but as signs of major social problems to come. A new gaze was directed at family life and children's behavior, a gaze shaped and educated in psychological terms. It was now not a matter of visualizing the family regime in terms of sobriety, diligence, and thrift, nor cleanliness, healthy diet, and hygiene. A psychological realm of intimacy was now visualized, made up of fears, early experiences, anxieties, attitudes, relationships, conflicts, feelings of persecutions, wishes, desires, phantasies, and guilt.

The language of mental hygiene made it possible to conceptualize a range of new institutions such as child guidance clinics, as the fulcrum of a comprehensive system for the inspection and treatment of all those pathologies now described as 'maladjustment'. It made it possible to reconceptualize existing institutions, from the courtroom to the factory, in terms of mental hygiene, seeing problems within them in terms of poor mental hygiene and defining a future for them in terms of their potential for promoting health, contentment, and efficiency. The new language was disseminated to family members through the broadcasts and popular writings of psychologists and experts on child development. Mental hygiene was both a 'public' and a 'private' value: it linked social tranquillity and institutional efficiency with personal contentment. In providing the vocabulary for thinking and talking about human development and human troubles in psychological terms, in promoting the scrutiny and regulation of family life and institutional existence in terms of mental normality, psychology linked subjective and intersubjective existence into governmental programs in a new way.

In wartime and the postwar period, a new set of psychological terms entered the language of government. This new language of 'the group' opened the relational life of organizations up to thought and action in a new way (Miller and Rose, 1988; Rose, 1989a; see this volume, Chapter 6). The language of the group made intelligible a range of problematic phenomena – from the unmanageable factory to the discontented battalion – that did not seem intelligible in terms of the characteristics of individuals. It amalgamated diverse and even apparently incompatible conceptual developments – object relations theory, Bion's concepts of the leaderless group, Kurt Lewin's field theory – and the notion of sociotechnical systems (Bion, 1961; Lewin, 1948). It became central to the postwar management of economic life, opening up the minutiae of interpersonal relations within the enterprise to analysis and interventions that could answer, simultaneously, to psychological principles of health and managerial principles of productivity. Psychological expertise was to be a crucial mediating element not only between employee and employer, but also between the desire of politicians for increased production and the private ownership and control of the production process.

Such associations between governmental ambitions, organizational demands, scientific knowledge, professional expertise, and individual aspirations are fundamental to the political organization of liberal democracies. It is not only that regulation extends way beyond the control of the pathological persons and conditions and embraces, as its preferred mode of operation, the production of normality itself. Crucially, regulation does not take the form of the extension of direct state scrutiny and control into all the petty details of social, institutional, and personal life. Political authorities 'act at a distance' upon the aims and aspirations of individuals, families, and organizations. Such action at a distance is made possible by the dissemination of vocabularies for understanding and interpreting one's life and one's actions, vocabularies that are authoritative because they derive from the rational discourses of science, not the arbitrary values of politics. It depends upon the accreditation of experts, who are accorded powers to prescribe ways of acting in the light of truth, not political interest. And it operates not through coercion but through persuasion, not through the fear produced by threats but through the tensions generated in the discrepancy between how life is and how much better one thinks it could be.

Inscribing subjectivity

For a domain to be governable, one not only needs the language to render it into thought, one also needs the information to assess its condition (Rose, 1988; Latour, 1986b; cf. Lynch and Woolgar, 1988). Information establishes a relay between authorities and events and persons at a distance from them. It enables the features of the domain accorded pertinence – types of goods and labor, ages of persons, prevalence of disease, rates of birth or death – to be represented in a calculable form in the place where decisions are to be made about them, in the manager's office, the war room, the case conference, the ministry for economic affairs, and other such *centers of calculation.* Projects for the government of social life were dependent upon the invention of devices for the inscription of subjectivity.

Statistics – literally the science of state – was originally the project to transcribe the attributes of the population into a form where they could enter into the calculations of rulers (Cullen, 1975; Hacking, 1986). The projects of police in the eighteenth century inspired a huge labor of inquiry. Inspectors were sent out and information gathered throughout the realm on the numbers of persons, their wealth and forms of habitation and trade, their births, illnesses, and deaths. From this cascade of figures, transported and communicated from all corners of the land, tables were compiled, charts drawn up, rates calculated, trends noted, averages compared, changes over time discovered. This was the technical dimension of the 'discovery of society' to which I have already referred: these numbers would henceforth be seen as indispensable to the question of the exercise of government by authorities, whether these be politicians, professionals, or philanthropists.

The transformation of the population into numbers that could be utilized in political and administrative debates and calculations was extended into a statisticalization of the morals and pathologies of the population. The statistical societies in Britain compiled charts and tables of domestic arrangements, types of employment, diet, and degrees of poverty and want (Abrams, 1968). Moral topographies of the population were constructed, mapping pauperism, delinquency, crime, and insanity across space and time and drawing all sorts of conclusions about changing rates of pathology, the causes of such conditions, and the measures needed to alleviate them (Jones and Williamson, 1979; Rose, 1979). As statistics were gathered about pathologies of the moral order, these too appeared to manifest lawlike characteristics. It now began to be argued that there were laws of the moral order that gave rise to regularities in individual conduct (cf. Hacking, 1990). The oft repeated question of the relation of society to the individual began to take the form that would be recognizable for the next one hundred years.

These, however, were superficial measures of subjectivity, relying on counting cases of evident pathological conduct after they became known to the authorities. Psychology would stake its claim to be an effective social discipline on its capacity to individualize subjects in a more fundamental manner. Michel Foucault argued that all the sciences which have the prefix psy or psycho have their roots in a transformed relationship between social power and the human body, in which regulatory systems sought to codify, calculate, supervise, and maximize the levels of functioning of individuals (Foucault, 1977). In this 'reversal of the political axis of individualization', the kinds of detailed attention that had previously been focused only on the privileged – royalty, nobility, the wealthy, the artist – now came to be directed upon the infamous, notably the criminal, the lunatic, the pervert, and the schoolchild.

The power of psychology here lay in its promise to provide inscription devices that would individualize such troublesome subjects, rendering the human soul into thought in the form of calculable traces. Its contribution lay in the invention of diagnostic categories, evaluations, assessments, and tests that constructed the subjective in a form in which it could be represented in classifications, in figures and quotients. The psychological test was the first such device. Codification, mathematization, and standardization make the test a minilaboratory for the inscription of difference, enabling the realization of almost any psychological scheme for differentiating individuals in a brief time, in a manageable space, and at the will of the expert (Rose, 1988). Tests and examinations combine power, truth, and subjectification: they render individuals into knowledge as objects of a hierarchical and normative gaze, making it possible to qualify, to classify, and to punish (Foucault, 1977, pp. 184–5). Hence the ritual of the test, in all its forms and varieties, has become central to our modern techniques for governing human individuality, evaluating potential recruits to the army, providing 'vocational guidance', assessing maladjusted children – indeed in all the practices where decisions are to be made by authorities about the destiny of subjects.

The translation of the individual into the domain of knowledge makes it possible to govern subjectivity according to norms claiming the status of science, by professionals grounding their authority in an esoteric but objective knowledge. In the school, factory, prison, and army, psychologists were to become experts on the rational utilization of the human factor. Psychology began to claim a capacity not merely to individualize, and classify, but also to advise upon all facets of institutional life, to increase efficiency and satisfaction, productivity and contentment.

It was here that the social psychologies were to be instated, playing a key role in rendering interpersonal relations thinkable and inscribable (these issues are discussed in more detail in Rose, 1989a, included in this volume as Chapter 6). Persons, it appeared, were not automata; they acted in terms of a subjective world of meanings and values, in short, of 'attitudes'. The concept of attitude went hand in hand with a method for inscribing it. The 'attitude survey' became a key device for charting the subjective world, enabling it to be turned into numbers and utilized in formulating arguments and strategies for the company, the political party, the military – indeed anywhere that individuals were to be governed through their consent. This psychological gaze was to be directed at the nation as a whole through such devices as public opinion surveys. The social psychology of opinions and attitudes presented itself as a continuous relay between authorities and citizens. Government needed to be undertaken in the light of a knowledge of the subjective states of citizens and needed to act upon that subjective state, if citizens were freely to discharge their social obligations. Through its capacity to inscribe and translate subjectivity, to provide a technology that would link up the will of citizens with the decisions of authorities, such a social psychology was to portray itself as no less than a science of democracy.

Technologies of subjectivity

It is thus no accident that psychology – as a language, a set of norms, a body of values, an assortment of techniques, a plethora of experts – plays such a significant role in technologies of government within liberal democracies. Such societies do not exercise power through the domination of subjects, coercing them into action by more or less explicit threats or inducements offered by the central powers (Miller, 1987). On the contrary, they establish a necessary distance between the legal and penal powers of the state and the activities of individuals. Government is achieved through educating citizens, in their professional roles and in their personal lives – in the languages by which they interpret their experiences, the norms by which they should evaluate them, the techniques by which they should seek to improve them. It is exercised through assemblages of diverse forces – laws, buildings, professions, techniques, commodities, public representations, centers of calculation, and types of judgment – bound into those more or less stable assemblages of persons, things, devices, and forms of knowledge that we term education, psychiatry, management, family life.

'The state' is neither the origin nor puppet master of all these programs of government. Innovations in government have usually been made, not in response to grand threats to the state, but in the attempt to manage local, petty, and even marginal problems. Programs for enhancing or changing the ways in which authorities should think about or deal with this or that trouble have sometimes issued from the central political apparatus, but more characteristically they have been formulated by lawyers, psychiatrists, criminologists, feminists, social workers, bosses, workers, parents. Effecting these programs has sometimes involved legislation, and sometimes entailed setting up new branches of the political apparatus, but it has also been the work of dispersed professional groups, voluntary, philanthropic, or charitable organizations. Innovations have been sporadic, often involving the ad hoc utilization, combination, and extension of existing explanatory frameworks and techniques. Some have come to nothing, failed, or been abandoned or outflanked. Others have flourished, spread to other locales and problems, established themselves as lasting procedures of thought and action. In the apparatuses and relations that have solidified, the very realities of state, politics, and society have been radically transformed.

Psychology is not merely a space in which outside forces have been played out, or a tool to be used by pregiven classes or interest groups. To the extent that various of its theories have been more or less successful in enrolling allies in their support, in producing calculable transformations in the social world, in linking themselves into stable social networks, they have established new possibilities for action and control. In establishing and consolidating such networks, in forcing others to move along particular channels of thinking and acting, psychologists have participated in the fabrication of contemporary reality. It is not merely that new ways have been introduced for construing the entities and relations that exist in the world. It is also, as we have seen, that new associations have been established between a variety of agents, each of whose powers may be enhanced to the extent that they can 'translate' the arguments or artifacts in question so that they may function to advantage in relation to their particular concerns or ambitions (cf. Latour, 1984, 1986a; Callon, 1986; Law, 1986, 1987). Technologies of government, that is to say, take the form of loose associations linking diverse agents through a series of relays through which the objectives and aspirations of those at one point – departments of state, expert committees, professionals, managers – can be translated into the calculations and actions of those distant from them in space and time, such as health visitors, teachers, workers, parents, and citizens.

Psychology has become basic to such associations. It provides the languages to establish translatability between politicians, lawyers, managers, bureaucrats, professionals, businessmen, and each of us. It establishes the norms and techniques that can be applied in so many different contexts. And its expertise, grounded not in political partiality but in a claim to truth based on knowledge and training, allows for an indirect relationship to be established between the ambitions of governmental programs for mental health, law and order, industrial efficiency, marital harmony, childhood adjustment, and the like, and the hopes, wishes, and anxieties of individuals and families. Convinced that we should construe our lives in psychological terms of adjustment, fulfillment, good relationships, self-actualization, and so forth, we have tied ourselves 'voluntarily' to the knowledges that experts profess, and to their promises to assist us in the personal quests for happiness that we 'freely' undertake.

The soul of the citizen

Psychology, then, has been bound up with the entry of the soul of the citizen into the sphere of government (Gordon, 1987; Rose, 1990). The apparently 'public' issue of rationalities of government is fundamentally linked to the apparently 'private' question of how we should behave, how we should regulate our own conduct, how we should judge our behavior and that of others. This link has not been a merely 'external' one, in which government has sought to manipulate otherwise 'free' individuals. It has been an 'internal' one, in which our very constitution as 'free' individuals has been the objective and consequence of regulatory programs and techniques (Foucault, 1982; Rose, 1993).

As early as doctrines of police, an explicit relationship was established between government of a territory and government of oneself. The individual was to be taught "to control his own life by mastering his emotions and to subordinate himself politically without resistance" (Oestreich, 1982, p. 164). This entailed a training in the minute arts of self-scrutiny, self-evaluation, and self-regulation ranging from the control of the body, speech and movement in school, through the mental drill inculcated in school and university, to the Christian practices of self-inspection and obedience to divine reason. Only to the extent that such self-regulatory practices were installed in subjects did it become possible to dismantle the mass of detailed prescriptions and prohibitions concerning the minutiae of conduct, maintaining them only in limited and specialized institutions: penitentiaries, workhouses, schools, reformatories, and factories. Through such practices of the self, individuals were to be subjected not by an alien gaze but through a reflexive hermeneutics.

The concept of subjection may suggest that persons are entrapped in devices whose ends they do not share. Of course, there is an important sense in which this was and is the case in moralizing machines such as prisons, the army, the factory, the school, and even the family. But even in the nineteenth-century prison, the aim of the isolation, the daily delivery of moral injunctions from the pulpit, and the reading of the Bible was to provoke self-reflection on the part of the inmate. The rationale was to transform the individual not merely through mindless inculcation of habits of obedience, but through the evoking of conscience and the wish to make amends. In the classroom, the factory, and the asylum ward, the moral order of the child, the laborer, or the lunatic was to be restructured such that the individual would take into himself or herself the constant judgment of skill, punctuality, comportment, language, and conduct that were embodied in the organization and norms of the institution.

Thus these apparatuses did not seek to *crush* subjectivity but to produce individuals who attributed a certain kind of moral subjectivity to themselves and who evaluated and reformed themselves according to its norms. This should not be viewed in terms of ideology but analyzed as what Foucault termed *technologies of the self,* "which permit individuals to effect by their own means or with the help of others a number of operations on their own bodies and souls, thoughts, conduct and way of being, so as to transform themselves in order to attain a certain state of happiness, purity, wisdom, perfection or immortality" (Foucault, 1988, p. 18).

If psychology has played a key role in the technologies that produce the modern subject as a 'self' of a certain type, this has not been merely through its individualistic, adaptive, and behaviorist branches. For in contemporary rationalities and technologies of government, the citizen is construed and addressed as a subject actively engaged in thinking, wanting, feeling, and doing, interacting with others in terms of these psychological forces and being affected by the relations that others have with them. It is upon these social and dynamic relations that government seeks to act. In the family, the factory, and the expanding systems of counseling and therapy, the vocabularies of mental hygiene, group relations, and psychodynamics are translated into techniques of self-inspection and self-rectification. These techniques are taught by teachers, managers, health visitors, social workers, and doctors. Through the pronouncement of experts in print, on television, in radio phone-ins, they are woven into the fabric of our everyday experience, our aspirations and dissatisfactions. Through our attachment to such technologies of the self, we are governed by our active engagement in the search for a form of existence that is at once personally fulfilling and beneficial to our families, our communities, and the collective well-being of the nation.

Within contemporary political rationalities and technologies of government, the freedom of subjects is more than merely an ideology. Subjects are *obliged* to be 'free', to construe their existence as the outcome of choices that they make among a plurality of alternatives (Meyer, 1986). Family life, parenting, even work itself are no longer to be constraints upon freedom and autonomy: they are to be essential elements in the path to self-fulfillment. Styles of living are to be assembled by choice among a plurality of alternatives, each of which is to be legitimated in terms of a personal choice. The modern self is impelled to make life meaningful through the search for happiness and self-realization in his or her individual biography: the ethics of subjectivity are inextricably locked into the procedures of power.

Modern citizens are thus not incessantly dominated, repressed, or colonized by power (although, of course, domination and repression play their part in particular practices and sectors) but subjectified, educated, and solicited into a loose and flexible alliance between personal interpretations and ambitions and institutionally or socially valued ways of living. The languages and techniques of psychology provide vital relays between contemporary government and the ethical technologies by which modern individuals come to govern their own lives. They are increasingly purveyed, not by univocal moralistic interventions of social agencies but through multiple voices of humanistic and concerned professionals, whose expert advice on the arts of existence is disseminated by the mass media. These may be polyvocal but they offer us solutions to the same problem – that of living our lives according to a norm of autonomy. Their values and procedures free techniques of self-regulation from their disciplinary and moralistic residues, emphasizing that work on the self and its relations to others is in the interests of personal development and must be an individual commitment. They provide languages of self-interpretation, criteria for self-evaluation, and technologies for self-rectification that render existence into thought as a profoundly psychological affair and make our self-government a matter of our choice and our freedom. And for those selves unable to conform to the obligations of the free subject, unable to choose or anguished by the choices they have made, dynamic and social therapies offer technologies of reformation consonant with the same political principles, institutional demands, and personal ideals. They are mainly supplied by free choice in the market. They are legitimated in terms of their truth or their efficacy rather than their morality. And they promise to restore the subject to autonomy and freedom. Government of the modern soul thus takes effect through the construction of a web of technologies for fabricating and maintaining the self-government of the citizen (Rose, 1990; Miller and Rose, 1990).

Genealogies of the subject

In the complex of powers over subjectivity entailed in modern apparatuses of regulation, 'the social' has traced out the very topography of our soul. We are governed through the delicate and minute infiltration of the dreams of authorities and the enthusiasms of expertise into our realities, our desires, and our visions of freedom. To write the genealogy of psychology in such terms is not, however, to subject it to a critique. Genealogy seeks not to reveal falsity but to describe the constitution of truths. It does not ask Why? but How? It does not simply reverse hierarchies – pure versus applied, soul versus body, ethics versus administration, social versus personal – but fragments them. It attends to the 'marginal' and shows its centrality, to the pathological as the condition for normality, to that considered inessential to show how, through it, the essential has been fabricated. And if the genealogy of psychology brings into focus the parts that orthodoxy considers impure and shameful, it does so not to denounce but to diagnose, as a necessary preliminary to the prescription of antidotes.

4

Expertise and the *techne* of psychology

Over the past half century, in the liberal, democratic, and capitalist societies of what we used to call the West, the stewardship of human conduct has become an intrinsically psychological activity. Psychological experts, psychological vocabularies, psychological evaluations, and psychological techniques have made themselves indispensable in the workplace and the marketplace, in the electoral process and the business of politics, in family life and sexuality, in pedagogy and child rearing, in the apparatus of law and punishment, and in the medico-welfare complex. Further, it is increasingly to psychologists that the citizens of such societies look when they seek to comprehend and surmount the problems that beset the human condition – despair, loss, tragedy, conflict – living their lives according to a psychological ethic. The rise of 'the psychological' is thus a phenomenon of considerable importance in attempting to understand the forms of life that we inhabit at the close of the twentieth century.

A number of authors have documented the expansion of the psychological domain, and have accounted for it in different ways (e.g., Lash, 1979, 1984; Bourdieu, 1984; Baritz, 1960; M. Rose, 1978; Rieff, 1966). Whatever the undoubted strengths of these accounts, they have tended to see psychology – and the psy knowledges and devices more generally – as little more than *signs* or *effects* of other, more fundamental, social or cultural changes, and indeed have tended to view these events as *symptoms* of a more general cultural malaise. However, I wish to suggest that the kinds of phenomena that I loosely describe as 'psy' have a more significant role than this, and to propose some lines for the investigation of the mechanisms by which psychology has achieved such a wide penetration into modern life in the West. In doing so, my aim is not to bring forward new evidence, but rather to suggest some ways of understanding the place of these 'engineers of the human soul' within the sociopolitical arrangements of liberal democratic societies.

Such an investigation has a particular salience as the twentieth century draws to a close. The societies of Eastern Europe are attempting to cast off their allegiance to the political problematics of Marxism-Leninism and to the associated regulatory technologies of the party apparatus, the central plan, and the ethics of social duty and collective responsibility. In their place, they look to the economic and industrial technologies and expertise of the West in order to reconstruct their economic orders on the principles of the market, competition, and enterprise. But the experience of the West might imply that there is also a relationship between liberal democracies and expert technologies of a different sort. This would be a relationship between political problematics articulated in terms of individualism and freedom and the expertise of the psy disciplines. It is this integral relationship between individuality and knowledge, between freedom and expertise that I explore here.

Psychological applications?

How should one study the 'social' role of psychology, the actual social existence and operation of its truths, procedures, and personnel and their linkages to the academic field of the discipline? Perhaps my own perspective upon these issues is best introduced by distinguishing it, rather rapidly and at the risk of some oversimplification, from three other approaches: histories of applications, histories of ideas, and histories of professions.

Let me begin with the issue of 'application'. In the histories of psychology written by psychologists themselves, and indeed frequently in those written by others, the social role of psychology is characterized through the notion of application. In such accounts, psychology is portrayed as a fortunate science, in that the discoveries that psychologists make in their cogitation and research, and the techniques that they devise in the academy and the psychological laboratory can be put to use in the resolution of problems in the world outside. Such an analytic of application is, however, misleading. It suggests that the line of development of a discipline like psychology runs from the pure to the applied, from the normal to the pathological, from the laboratory to society. In fact, as I and others have argued, it is more enlightening to pose the question the other way around (Rose, 1985a; cf. Danziger, 1990; Smith, 1988; Canguilhem, 1978; Foucault, 1977). First, 'psychology' does not define by right and throughout history a coherent and individuated body of phenomena and explanatory projects existing in a continuous cognitive space. Psychology was only able to differentiate from medicine, philosophy, and physiology, only able to 'disciplinize' itself, on the basis of its social vocation, its elaboration within the educational, penal, military, and industrial apparatuses. It was by means of attempts to grasp and rectify conditions deemed pathological – feeblemindedness, industrial inefficiency, shell shock, maladjustment – that a psychological discipline of the normal individual appeared possible. The vectors of force can more usefully be traced not from 'ideas' to 'applications' but from problems and solutions to theories and experiments. The social role of psychology should not, therefore, be analyzed as a history of 'applications' but as a history of 'problematizations': the kinds of problems to which psychological 'know-how' has come to appear as solutions and, reciprocally, the kinds of issues that psychological ways of seeing and calculating have rendered problematic.

Psychological ideas?

If the relations of psychology to society are not captured in the notion of 'application', should they then be investigated in terms of the 'history of ideas'? A history of psychology conducted in these terms would trace the ways in which the concepts, concerns, metaphors, and types of explanation within psychology have resonances both with those that precede them and with themes in the writings in philosophy, literature, theology, and elsewhere with which they are contemporary. Such histories of psychology examine the consonance between prevailing cultural repertoires of explanation and organization and characteristic metaphors within psychological explanations, such as those of energy-based mechanisms in Freud or those of information flows in contemporary cognitive science. And they trace the ways in which psychological theories and explanations influenced – and were influenced by – those in other areas of society, pursuing these through the commonalities of biography and education, of class and gender backgrounds, and the personal relations between psychologists and other socially significant figures.

No doubt psychology is, in part, a matter of ideas, and its spread is, in part to be accounted for in terms of cultural resonances and personal influences. However, psychological ideas should, perhaps, be seen less as 'ways of thinking' than as 'intellectual techniques', ways of making the world thinkable and practicable in certain ways. 'Ideas' here are bound into ways of seeing and acting: into technologies. They are enmeshed in definite practices of experimentation, investigation, and interrogation arising not only in the laboratory or the academic's study but in an array of social locales. They circulate within particular apparatuses for the dissemination and adjudication of psychological truths – learned journals, university courses, conferences, symposia, newspaper articles, and books. They are engaged within definite forms of discourse having the character of investigation, diagnosis, inquisition, confession, or judgment. They are embedded in apparatus and technical devices, not only those of the laboratory, but also those of the assessment, the interview, the diagnosis, the counseling session, and the like. And they are deployed within a range of strategies and programs in which social authorities have construed their difficulties or their objectives as linked, in different ways, to the subjective capacities of humans and have called upon psychology and psychologists for assistance. To understand these events, I suggest, is neither a matter of burrowing through the minutiae of biography nor of gesturing to grand cultural affinities or world views, but of documenting the often mundane and marginal alliances, complicities and takeovers through which psychological ways of thinking and acting have ramified along these multiple pathways.

Psychological professionals?

A third way of thinking about the relations between psychology and society, and the processes through which psychology has inserted itself so insistently into our modern reality, focuses on the occupational strategies of psychologists themselves, analyzed by means of the notion of 'professionalization'. In its modern version, the notion of professionalization accounts for the expansion of a field like psychology in terms of the strategies that particular occupational sectors have adopted in order to capture and monopolize a region of the market (e.g., Baritz, 1960; Ewen, 1976; cf. Freidson, 1970, 1986; Scull, 1989). Scientists here are viewed as interest groups seeking to further their own financial and moral status by securing exclusive control over the terms in which particular social concerns are framed, and over the loci of power within various social apparatuses – of healing, of justice, of punishment, of pedagogy, and the like. Such critical analyses of professionalization have pointed to the ways in which professionals legitimate their social powers through their claim to possess esoteric knowledge and technical capacities not available to others. They have tried to demonstrate the ways in which the interests of professionals in expanding their domain have allied with the interests of other powerful social groups – politicians, the business classes, men endeavoring to confine women to the domestic sphere. And they have portrayed this process in a critical light, not only because of the powers it serves and the interests it legitimates, but, frequently, on account of the falsity of the claims of experts to have special technical capacities that would enable them to resolve the moral, political aspects of the issues over which they preside.

Although these arguments concerning professionalization are certainly suggestive, they sidestep a number of issues that are of considerable importance in trying to understand the social role and calling of psychology. First, the concept of professionalization tends to divert our attention from an analysis of the conditions under which a distinct *psychological* vocation took shape. How did problems come to be formed – in industry, the army, the family, the penal system – in a way amenable to 'subjectification', so that psychology and psychologists would be able to claim to provide solutions. Under what conditions did those engaging with the reform of conduct in diverse locations come to see their interests as bound up with an identity as practitioners of a particular disciplinary base? And what accounts for an identification of problems and experts with *psychology,* a form of knowledge whose epistemological credentials have always been contested and whose links to power have been – until recently – both limited and tenuous?

The notion of professionalization also obscures a second set of problems. These concern the specific ways in which psychology is organized as a practice – as distinct from the practices associated with medicine, law, or religion. This is a question, first of all, of attention to what one might term the 'ethos' of psychology, the characteristic ways in which the authority and morality of the psychologist are manifested in psychological discourse (for this use of Aristotle's notion of ethos, see Summa, 1990). And it is also a question of attention to the *techne* of psychology, the ways in which it is organized as a practice – in terms of the locales within which psychologists work, the varied techniques available to them, their diverse modes of operation, the very different relations of authority, domination, and alliance between psychologists and their subjects.

Perhaps most significantly for the present argument, the notion of professionalization does not tell us much about why *psychological* expertise has come to play such an important part within particular types of political and social thought associated with *liberal democratic* societies. There are, of course, at least two provocative accounts that link the success of occupational strategies of those professing neutrality, humanity, and efficacy to the nature of contemporary social relations – those offered by Talcott Parsons and Jürgen Habermas. Parsons, writing specifically of doctors, sees the characteristic ethical, technical, and vocational rights of the medical professional as reciprocally linked to a social system in which deviance has come to be expressed through a kind of motivated withdrawal of the individual into the sick role of patient, equipped with a condition that the doctor must simultaneously ratify and redress in the interests of the functional prerequisites of the social system (e.g., Parsons, 1939, 1951a, 1951b). But the rise of specifically psychological ways of thinking and acting is elsewhere viewed by Parsons as largely an effect: a response to prior structural and systemic changes in society that have placed greater strains on the psychological functioning of individuals in families, work, and elsewhere. This perspective does not allow much scope for an investigation of the particularities of the particular profession of psy and its active role in shaping the world of which it now forms a part.

On the other hand, Habermas suggests that the proliferation of professional activities might be understood in terms of the expansion of purposive-rational action, with a consequent colonization of the lifeworld, an expansion of formal rational knowledge and system rationalization, the domination of scientific and technological rationality over other value spheres (Habermas, 1971, 1972, 1984). Thus he argues that the spontaneous processes of everyday practice have been deformed by an expert culture embodied in a "dense network of legal norms, of governmental and paragovernmental bureaucracies" (Habermas 1987, p. 361). However, as I have suggested elsewhere, these themes are at once too general and too pessimistic to analyze the rise of the expertise of human conduct (Rose, 1994b). Too general because they confer too much unity on the diversity of ways in which the truth claims, technical procedures, and authoritative powers of experts have helped make up our present. Too pessimistic because they apply too simple a formula for calibrating the transformation in the relations between experts, political power, and the management of the self in contemporary societies. This is particularly the case when such approaches are utilized to analyze the specific role of psy expertise in the strategies and technologies of government that I term 'advanced liberal', which have come to accord a particular value to notions of autonomy, choice, and the pursuit of personal identity.

Thus neither the account provided by Parsons nor that offered by Habermas appears to me to capture precisely what it is that has been so significant about *psychology* as a particular mode of speaking the truth about human subjects and a particular mode of acting efficaciously upon human subjects in order to improve them. What is it that is specific about psychology that accounts for the particular potency that it has in lending its coloration to so many of the practices, locales, and forms of judgment in the societies of the West?

Psychology as expertise

The 'social' role of psychology becomes clearer when we approach it not as a matter of appliance of science, not as a matter of the evolution of ideas, not even as a matter of the rise of a profession, but in terms of *expertise.* I use the term 'expertise' to refer to a particular kind of social *authority,* characteristically deployed around *problems,* exercising a certain *diagnostic* gaze, grounded in a claim to *truth,* asserting technical *efficacy,* and avowing *humane* ethical virtues. In advocating the use of the notion of expertise, rather than that of professional group or occupational sector, I mean to focus attention upon three characteristics. First, whereas the notion of professionalization implies an attempt to found occupational exclusiveness on the basis of a monopolization of an area of practice and the possession of an exclusive disciplinary base, expertise is heterogeneous. Its characteristic style of activity is that of *bricolage:* it amalgamates knowledges and techniques from different sources into a complex 'know-how'. Only later is the attempt made to ratify the coherence of this array of procedures and forms of thought, to formalize them as a certain 'specialism'. Even then, this characteristically proceeds not by seeking to derive them from a single body of theory but by unifying them within a pedagogic practice: courses, credentials, textbooks, professional qualifying examinations, and the like.

Second, the notion of expertise enables us to distinguish between a number of different processes: the occupational advancement of sectors professing a certain kind of knowledge; the capacity of certain forms of thought to render problems intelligible through inquisition, scrutiny, or calculation; the invention of particular practicable technical devices; the transformation of regulatory practices in the light of those truths and techniques. To put this rather crudely in relation to psychology, one would want to recognize that the social consequences of psychology are not the same as the social consequences of psycholog*ists*. Psychology is a 'generous' discipline: the key to the social penetration of psychology lies in its capacity to lend itself 'freely' to others who will 'borrow' it because of what it offers to them in the way of a justification and guide to action. It is in this fashion that psychological ways of thinking and acting have come to infuse the practices of other social actors such as doctors, social workers, managers, nurses, even accountants. Psychology enters into alliance with such agents of social authority, colonizing their ways of calculating and arguing with psychological vocabularies, reformulating their ways of explaining normality and pathology in psychological terms, giving their techniques a psychological coloration. Through such alliances psychology has made itself powerful: not so much because it has managed to exclude others from making psychological statements or psychological judgments, but because of what it has provided for others, on condition that they come to think and act, in certain respects, like psychologists.

Third, the notion of expertise enables us to grasp something rather significant about the ways in which such affiliations arise. For these alliances do not simply provide psychology with a means to gain its hold on social reality, as it were, by proxy. They also provide something for the doctors, nurses, social workers, and managers who enter into psychological coalitions. Those who enter such arrangements with psychology are those engaged in all the proliferating practices that have to deal with the vagaries of human conduct and human pathology, all those who seek to act upon it in a reasoned and calculated form. Psychology can graft itself onto such practices because it promises a certain 'simplification' of the heterogeneous tasks of authorities. It appears as if it can systematize and lend coherence to the ways in which authorities visualize, evaluate, and diagnose the conduct of their human subjects – render rational the grounds of decision and action. In purporting to underpin authority by a coherent intellectual and practical regime, psychology offers others both a grounding in truth and some formulas for efficacy. But further, and perhaps crucially, in claiming to modulate power through a knowledge of subjectivity, psychology can provide social authority with a basis that is not merely technical and scientific but 'ethical'.

The techne *of psychology*

Approached from the perspective of expertise, we are in a better position to connect up our analysis of the proliferation of psychology with a number of other reflections on transformations in social arrangements and forms of authority in European societies over the past century. From this perspective, our focus shifts from psychology itself to the modes in which psychological knowledges and techniques have grafted themselves onto other practices. Psychology here is seen as offering something to, and deriving something from, its capacity to enter into a number of diverse 'human technologies'.

The notion of a human technology is not intended to imply an *inhuman* technology – one that crushes and dehumanizes the essential personhood of those caught up within it. Quite the reverse: it is most frequently the promise of personhood, of being adequate to the real nature of the person to be governed, that underlies the power that psychology seeks and finds within such technologies. I use the term 'technology' to direct our analyses to the characteristic ways in which practices are organized to produce certain outcomes in terms of human conduct: reform, efficiency, education, cure, or virtue. I seek to draw attention to the complex technical forms invented to produce these outcomes – ways of combining persons, truths, judgments, devices, and actions into a stable, reproducible, and durable form. And I suggest that psychological modes of thought and action have come to underpin – and then to transform – a range of diverse practices for dealing with persons and conduct that were previously cognized and legitimated in other ways, via the charisma of the persona of the authority, by the repetition of traditional procedures, by appeal to extrinsic standards of morality, by rule of thumb.

Psychology, that is to say, does not principally find its sites of application within its own institutional locations – unlike the ancient professions, it has no church within which to redeem sin, no court of law within which to pronounce judgment, no hospital within which to diagnose or cure. Nonetheless it will find its social territory in all those proliferating encounters where human conduct is problematized in relation to ethical standards, social judgments, or individual pathology. What is it that psychology can offer to such encounters?

Calculable individuals

Calculability is a central theme in much sociological and philosophical reflection upon the 'uniqueness of the West', on the particular characteristics of the social arrangements and ethical systems that have emerged in the capitalist, bureaucratic, and democratic societies of northwest Europe and North America. Marx, Nietzsche, Weber, Lukacs, Habermas, and Foucault each, in his different way, suggested that calculation and calculability have become central not only in projects for the domination of nature, but also in relation to human beings. We have entered, it appears, the age of the calculable person, the person whose individuality is no longer ineffable, unique, and beyond knowledge, but can be known, mapped, calibrated, evaluated, quantified, predicted, and managed.

For those who take their cue from Marx, it is in the workplace and in the activity of production that the rise of calculability is to be grounded: F. W. Taylor's *Principles of Scientific Management* is seen as a response to the capitalist imperative of management, prediction, and control of labor, and psychotechnics as merely giving this imperative of calculability a pseudolegitimate form (Taylor, 1913). For those who take their cue from Weber, calculation is an inherent part of rational administration, bound up with the desire for exactitude, predictability, and the subordination of substantive or ad hoc judgment to the uniformity of a rule (Weber, 1978, chap. 11). In each of these cases, the calculability of the person is seen as the effect or symptom of a process that has its roots elsewhere. But perhaps we should turn away from such sweeping characterizations of social epochs, to ask some more modest questions. How – under what practical conditions and within what social arrangements – was it possible for the human individual to become calculable? How – through what procedures of inscription, differentiation, and cognition – did the knowledges and procedures emerge that would make of the human being a calculable entity? How did this calculation come to appear a matter not of arbitrary decisions, particular value choices, or disputable social goals, but of objective criteria, arising out of scientific investigation of the nature of indisputable processes, and made through procedures that are not political but technical?

The social vocation of psychology and its status as expertise is intrinsically bound to such questions. For it was through the formation of a specifically psychological expertise, and through the construction of institutional technologies that were infused by specifically psychological values, that individual difference became scientifically calculable and technically administrable (Rose, 1985a, 1988; see this volume, Chapter 5). To understand the birth of this *techne,* we need to adopt what Colin Gordon has aptly termed an 'institutional epistemology': that is to say, a perspective that views knowledges as arising out of the practical, technical organization of space, time, bodies, and gazes (Gordon, 1987). A psychological knowledge of individual differences did not emerge from a mysterious leap of the intellect or from laborious theoretical and scientific inquiry, but neither did it merely answer to the demand that capitalist control of the labor process be legitimated, or commend itself because of its elective affinity with a rationally calculating 'spirit of the age'. Rather, it needs to be understood in relation to the mundane organizational practices of those social apparatuses constructed in so many European states in the late nineteenth century that sought to organize persons en masse in relation to particular objectives – reform, education, cure, virtue. Schools, hospitals, prisons, reformatories, and factories acted as apparatuses for the isolation, intensification, and inscription of human difference. They played not only the part of microscope but also that of laboratory, for they were simultaneously locales of observation of and experimentation with human difference. Knowledge, here, needs to be understood as itself, in a crucial sense, a matter of technique, rooted in attempts to organize experience according to certain values.

Truth thus takes a technical form; truth becomes effective to the extent that it is embodied in technique. The paradigmatic technique of the calculable person is the psychological 'test'. The test, in all its forms, is a device for visualizing and inscribing individual difference in a calculable form. As Michel Foucault pointed out, tests and examinations combine power, truth, and subjectification: they render individuals into knowledge as objects of a hierarchical and normative gaze, making it possible to qualify, to classify, and to punish (Foucault, 1977, pp. 184–5). The diagnostic procedures associated with the test enable the invisible, subjective world of the individual to be visualized and represented in classifications, in figures and quotients. The test becomes a tiny but highly transferable diagram of a procedure for the inscription of human difference into the calculations of authorities. The psychological test thus plays a crucial organizational role within the *techne* of calculation that has become central to all those 'disciplinary' institutions that are the other 'normalizing' side of liberal democracy, institutions where individuals are to be governed in terms of their individuality, in such a way as to maximize their organizational utility and utilize their powers in a calculated form.

The critics of psychology often portray its utilization of tests and numbers as the antithesis of humanity and democracy, berating those who advocate or use them for their tendency to reduce the person to a mere number, a label, a fixed index of human worth and human inequality. But perhaps it is as instructive to attend to the opposite. One dimension of the 'power of psychology' lies precisely in the new *democratic* legitimacy that science accords to judgments of human difference. Psychology is potent because it can appear to shift such judgments from a sphere of values, prejudice, or rule of thumb to the sphere of human truths, equality of standards, cogently justifiable choices and objective criteria of efficacy that should reign in a democracy (cf. Miller and O'Leary, 1987, 1989; Miller and Rose, 1995). Psychological expertise renders human difference *technical.* A wealth of cogent criticism has been directed at one particular version of the psychological test – the IQ test, with its associated baggage of assumptions concerning intelligence and heritability. But it is nonetheless the case that the claim of psy testing to render the person objectively calculable, to differentiate according to nature and not according to prejudice, has made the spread of the psychological *techne* of calculation more generally (personality tests, aptitude tests, tests for vocational guidance, and job selection) both possible and justifiable. If psychological calculation and its associated expertise have infused all those bureaucratic practices for the distribution of persons to diverse 'treatments' (different schools, different classes, different punishments, different battalions) or to diverse roles within disciplinary organizations (tasks in the factory, ranks in the army, streams in the school), it is because of its simultaneous claim to efficacy and to humanity, to answer to the demands only of natural differences and human truths.

Manageable spaces

Psychology is often criticized for its individualism. But it would be wrong to think that, by this token, its sole contribution to our reality was to the techniques of individualization and the administration of individuals. Psychology also makes possible a *techne* of spaces and relations, coming to infuse all those practices where authorities have to administer individuals in their collective existence. Space, here, should not be understood as a primordial given, an a priori of thought or a straightforward matter of topography. Psychology is involved in the invention of spaces, in the opening up of certain fields for thought and action that simultaneously impart a psychological character to such spaces and enable them to be administered in the light of this character. What is it that psychology offers to those charged with the administration of collective activity?

First, again, calculability. Psychological expertise makes intersubjectivity calculable, enabling the calculated supervision and administration of collectivities. The social space of the factory, the school, the organization, and indeed the 'public' sphere itself has thereby been opened to calibration and management (Rose, 1989a; Miller, 1992). Whether via the notions of individual attitudes, of public opinion, of the human relations of the workplace, of the psychodynamic relations of the organization – those who are charged with the responsibility for administering the social existence of individuals may redefine their task in psychological terms. The language and techniques of attitude measurement and the attitude survey open the social actions of individuals to systematic planning and management by authorities. The notion of the group opens a human field for thought, argument, and administration that inhabits the architectural and social space of the factory, the schoolroom, the hospital, and the office (Miller and Rose, 1988, 1994; in this volume, see Chapter 6). The notion of public opinion and the technique of the opinion poll open a relation between political authorities and those they govern that goes beyond the periodic subjection of political leaders to a democratic mandate that takes the form of mere acclamation or condemnation. In each case, the dream is that psychological expertise can produce techniques that will allow the continuous adjustment of the decisions taken by authorities, or the type of justification offered for such decisions, in the light of the subjective commitments, values, and motivations of those over whom authority is to be exercised. Through such a process, authority is to be made both effective and legitimate to the extent that it is exercised in the light of a knowledge of those who are governed.

Ethicalizing authority

But let us reverse our gaze, from the field over which authority is to be exercised to the authoritative personages themselves. Psychological expertise not only promises a kind of *techne* for the administration of persons in their intersubjective relations. It also promises something to those who have the responsibility of wielding a power over others. On the one hand it enables them to assemble their various tasks and activities within a certain order and to subject them to a consistent set of calculations. On the other, it promises to 'ethicalize' the powers of authorities, from business consultants in search of profits and harmony to military men in search of efficient fighting forces. Through composing this image of rationality and ethicality, psychological expertise can promise to make authority simultaneously artful and wholesome.

No doubt for many centuries authorities have sought justification for the exercise of their power – legal, theological, moral, medical. Yet psychology has a characteristic and seductive relation to the practices it has come to regulate, in that it offers a means of exercising power that is ethical because it has as its basis not an external truth – be it divine right or collective good – but an internal truth, one essential to each individual person over whom it is exercised. Psychology grafts itself onto practices of law, punishment, management, parenthood, through its promise to combine efficacy and utility with humanity and fidelity. In establishing this relationship of guidance and reliance with the manager, the parent, the social worker, psychology turns the ideal personage of authority into a psychological calculator – one who visualizes the factory or the family in psychological terms, analyzes its strengths and weaknesses in psychological vocabularies, and makes decisions according to a psychological calculus. These provide those in authority with a way of deliberating about, judging, organizing the multitude of decisions that confront them as to how they should shape the conduct of those over whom they exercise that authority. No longer are the various activities of the manager or the parent merely an array of tasks that happen to coincide upon the person of the decision maker. These tasks undergo a radical simplification. They become patterned and can be linked up, related, explicated in terms of knowledge, reached according to certain established formulas, and adjudicated in terms of justified criteria. Whether instructing, managing, curing, punishing, educating, or reforming, one can first 'understand' via a hermeneutics of the soul conducted in psychological terms, one can then 'diagnose' according to a cogently justifiable classificatory system, and finally one can 'prescribe' a response via a calculated knowledge of subjectivity and techniques for its transformation.

But it is not simply that psychological expertise is 'simplifying', for this, after all, could be said of any other form of expertise to the extent that it renders a diverse assemblage of issues cognizable within a single explanatory space. It is also that exercising mastery over others in the light of a knowledge of their inner nature makes authority almost a therapeutic activity (cf. Miller and Rose, 1994). To this extent, one might see the spread of psy as answering to something of a 'crisis' – not a crisis of identity or subjectivity brought about by a stage of the process of modernization, but a crisis of authority. By this I mean that it relates to a concern posed to and by those who claim authority over conduct – social workers, teachers, managers, prison officers, and so forth – as to the justification for their authority and the ways in which it can claim to be legitimate in a democracy which accords rights and the protections of due process to its citizens. With the infusion of psy into the training and credentializing of professionals of conduct, the possibility emerges that the decisions that are made by such authorities can be conducted in a way that appears to be in the best interests of those whose lives they will affect – be they worker, prisoner, patient, or child. This ethical-therapeutic transformation is one aspect of the force that bonds diverse authorities to psychological expertise and makes it so powerful. It also explains the seductive promise held out by psychology to those who will exercise authority. It is not merely that it 'professionalizes' them, by supplying them with a knowledge base and so forth. It gives a new kind of human and moral worth and legitimacy not merely to the gross and evident wielding of power over others, but also to the mundane activities of daily decision making in the factory or in the family.

Hence a new kind of relationship is established between the psychological experts and those who consult them. Whether they be managers, parents, or patients, their relation to authority is a matter neither of subordination of will nor of rational persuasion. Rather, it has to do with a kind of discipleship. The relation between expert and client is structured by a hierarchy of wisdom, it is held in place by the wish for truth and certainty, and it offers the disciple the promise of self-understanding and self-improvement. It is not merely the promise of professional advancement that attracts business people, social workers, doctors, police officers, and so many others to psychologically informed training courses in managerial skills. Nor is it merely the hope that, once schooled in psychological vocabularies, techniques, and ways of calculating, one will be able simultaneously to do a good job and do good. The insight conferred by the psychologization of one's job is also an insight into oneself and one's life. For the allure of psychology is that the ethical pathway for authority is also an ethical pathway for the self.

The *techne* established within these relationships between psychological expertise and 'private' authority – such as that in the business enterprise and the family – has a general, political significance within liberal democratic techniques of government. Through these relations, such domains can be regulated by means of, rather than in spite of, their autonomy and responsibility. Thus, for example, psychological expertise of social and personal life has been disseminated by health visitors, family doctors, domestic science classes, radio and television programs, magazines, and advertisements. The norms and vocabularies elaborated confer a new visibility upon the workings of the family and new ways of identifying its malfunctions. Certain aspects of parental relations and responsibilities toward children become visible; certain notions of development, adjustment, and maladjustment are used to judge them; certain ways are promulgated by which family members learn to render their life into speech in the form of problems requiring solution. Now parents themselves can take on calculations and make judgments in these terms. And, when problems get too great for self-regulation, individuals and families themselves consult the experts to seek to overcome the anxiety formed in the gap between what they are and what they want to be. The 'private' domain of the family, and analogously that of the factory, can thus be normalized through the anxiety of its internal authorities – parents, managers – without breaching its formal autonomy: 'private' authority is bound into 'public' values by means of psychological expertise.

Psychology does not simply ally itself with authorities in private domains by promising to solve their problems. In 'applying itself' to such problems it transforms their terms. Industrial accidents become a matter of the interpersonal relations of the workplace. Profitability becomes a matter of releasing the self-actualizing potential of the work force. Naughty children become a matter of the emotional heritage of the parent's own childhood. Career advancement becomes a matter of self-confidence, and self-assertiveness. Reciprocally, failure as an employee, as a parent, as a human being becomes a matter of 'low self-esteem' (Cruikshank, 1993). Marketing becomes a matter of segmenting consumers by their psychological profiles, and advertising a matter of linking your product with the desires of those who must come to purchase it (Miller and Rose, 1996). Each of these problems now becomes inconceivable in other than psychological terms.

Further, the application of psychological expertise to a domain is itself problematizing; it generates new ways of construing existence as potentially problematic. Particularly significant is the proliferation of the notion of risk. Here the retrodirecton of the psychological gaze can identify problems *in potentia* and hence generate prophylactic strategies for their preemptive solution (Castel, 1991; cf. Hacking, 1990). Whereas pathology was a property of the concrete individual, risk is composed from a combination of factors that are not necessarily pathological in themselves – family background, parents' job record, type of housing, smoking, or drinking: as Castel argues, the "site of diagnostic synthesis" shifts from the individual under assessment to "a relationship constituted among the different expert assessments which make up the patient's dossier" (1991, p. 282). In this process, in the shift of problematization from pathology to risk, normality itself is rendered as the fragile outcome of the successful if inadvertent averting of risk. And a new role is opened for experts – that of identifying, recording, assessing risk factors in order to predict future pathology and take action to prevent it. Expertise can claim to turn chance and happenstance into certainty and predictability; alternatively, others can demand that it does so and berate it when it fails to prevent delinquency, child abuse, and so forth. Psy professionals can be allotted the role of ensuring the protection of the community and its inhabitants through the identification of the riskiness of individuals, actions, and forms of life. This is not a role that all experts have sought or embraced – although some evangelical psychologists have sought to convince others that it is possible to predict delinquency, childhood pathology, or physical or sexual child abuse from early signs picked up by tests and assessments when interpreted by the trained professional using scientific formulas. In other cases the obligation to become a risk identifier has no doubt been an unwelcome imposition, accompanied as it has been with blame for failure to prevent some unfortunate event that can no longer be represented simply as fate, misfortune, or accident (for some U.K. examples, see Carson, 1990; Ritchie, Dick, and Lingham, 1996; NHS Management Executive, 1993). For present purposes, what is significant is that, within the new logics of risk, professionals acquire the obligation to bring the future into the present and render it calculable. Hence the production and protection of normality can itself become an endeavor suffused with psychological calculation: normality is to be produced by a permanent modulation of deliberations and decisions by psychology in the light of a calculation of their riskiness.

Working on our selves

The final aspect of psychology as a technology that I want to discuss concerns the capacity of psychological languages and judgments to graft themselves into the ethical practices of individuals. By this I mean their ways of evaluating themselves in relation to what is true or false, good or bad, permitted or forbidden. Ethics here is understood in terms of specific 'techniques of the self', practices by which individuals seek to improve themselves and their lives and the aspirations and norms that guide them. Of course, many different authorities – theological, medical, political, military – in different historical periods and different social settings have advised, instructed, warned, and moralized concerning such procedures for shaping a life. They have defined many systems for conducting the minutiae of an existence – eating, copulating, defecating, sleeping, waking, observances of faith – in order to follow the way to truth, virtue, happiness, or salvation. And they have elaborated many practices, from ritual dances to sermons, from trances to confessions, to craft, shape, and sustain the desired kinds of self.

Many have commented upon the ways in which such practices for the interpretation and improvement of the self in advanced liberal societies have achieved a psychological coloration, operating according to psychological norms and in relation to psychological truths (Rieff, 1966; Lash, 1979; Rose, 1990). Psychological languages and evaluations have transformed the ways in which we construe and conduct our encounters with others – with our bosses, employees, workmates, wives, husbands, lovers, mothers, fathers, children, and friends. Each mode of encounter has been reconfigured in terms of personal feelings, desires, personalities, strivings, and fears. Psychological techniques have come to infuse, dominate, or displace theological, moral, bodily, dietary, and other regimens for bringing the self to virtue or happiness, and also those deployed for reconciling the self to tragedy or disappointment. And if the experts on hand to guide us through the conduct of our lives are not all psychologists, they are nonetheless increasingly trained by psychologists, deploy a psychological hermeneutics, utilize psychological explanatory systems, and recommend psychological measures of redress.

The ethical technologies deployed within this regime are, of course, heterogeneous. Nonetheless, it is, perhaps, the *techne* of the confessional that has been most versatile and most transferable, able to travel easily from one locale to another. For Michel Foucault, confession was the diagram of a particular form of power (Foucault, 1978b). The truthful rendering into speech of who one is and what one does – to one's parents, one's teachers, one's doctor, one's lover – was both identifying, in that it constructed a self in terms of a certain norm of identity, and subjectifying, in that one became a subject at the price of entering into a certain game of authority. Not only does confession in this sense characterize almost all of the proliferating systems of psychotherapy and counseling. It also provides a potent technical form that has come to install itself in a range of other practices where the conduct of personal life is at stake, from the doctor's surgery to the radio phone-in, from the social work interview to the frank interchange of lovers.

To speak the truth of one's feelings and desires, to 'share' them as the saying has it, is not merely a rendering audible of the inarticulate murmuring of the soul. In the very technical form of therapeutic procedures – whether it be in psychotherapy, Rogerian client-centered therapy, Perls's gestalt therapy, Janov's primal screaming, feminist therapy, or in counseling – the confessing subject is *identified.* The 'I' that speaks is to be – at least when 'insight' has been gained – identical with the 'I' whose feelings, wishes, anxieties, and fears are articulated. One becomes, at least *in potentia,* the subject of one's own narrative, and in the very act itself one is attached to the work of constructing an identity. And in the very same moment when the subject is affiliated to such an identity project, he or she is bound to the languages and norms of psychological expertise. For the words and rituals that govern these confessions are those prescribed by an authority, albeit one who has replaced the claims of god and religion with those of nature and the psyche.

Some contemporary psychologists interpret the outcome of such processes, in which individuals scrutinize, interpret, and speak about themselves in a psychological vocabulary, in terms of the 'social construction' of the person (e.g., Shotter and Gergen, 1989). But perhaps we would be wise to be agnostic about such claims, at least insofar as they are linked to a project of developing an 'alternative' psychology. From my perspective, it is not a question of discovering what people are, but of diagnosing what they take themselves to be, the criteria and standards by which they judge themselves, the ways in which they interpret their problems and problematize their existence, the authorities under whose aegis such problematizations are conducted – and their consequences. If we have become profoundly psychological beings, it would, then, be less that we have been equipped with a psychology than that we have come to think, judge, console, and reform ourselves according to psychological norms of truth.

Confession has been joined by a range of other psychotechnologies of the self. Most significant, perhaps, has been the rapid spread of behavioral techniques for teaching the arts of existence as social skills. These are no longer practiced largely as quasi-punitive methods for disciplining and reforming pathological persons – pedophiles, alcoholics – by a regime that tries to eliminate the undesirable behavior by associating the very thought of it with unpleasant experiences such as pain or nausea. Nor are they confined to the elimination of phobias and irrational fears. Now they are deployed in a whole range of locales, from assertion training for women to dietary control, from the 'empowering' of the disadvantaged young mother whose children are considered 'at risk' to the 'normalization' of the psychiatric patient to permit an appropriate presentation of self to 'the community' and to potential employers (Cruikshank, 1994; Baistow, 1995).

For present purposes, the details of these diverse practices of psychological reform are less significant than the mode of operation of expertise that is involved. It is not only that the truths of psychology have become connected to our practices of the self, with the notions that normality, autonomy, and personal success can be achieved through the engagement of the self in a psychological regime of therapeutic remodeling. It is also that a psychological ethics is intimately tied to the liberal aspirations of freedom, choice, and identity. Therapeutic ethics promises a system of values freed from the moral judgment of social authorities. Its norms answer not to an arbitrary moral or political code but only to the demands of our nature and our truth as human beings. And it seeks not to impose a new moral self upon us, but to free the self we truly are, to make it possible for us each to make a project of our own lives, to fulfill ourselves through the choices we make, and to shape our existence according to an ethics of autonomy.

Critics have tended to view the multiplication of therapeutics as a symptom of cultural malaise: of the pervasive individualism of modern Western culture; of the decline of religion and other transcendental systems for imparting meaning to quotidian existence; of the transformation of familial authority and the rise of narcissism; of the loss of the old solidities in a postmodern world in flux. But perhaps we might understand this differently if we focus upon the prominence given by contemporary psychological techniques to the norm of autonomy and the promise of achieving it under a regime of rational management of existence. This should be connected to the value attached, in contemporary rationalities of government, to the notions of individual liberty, choice, and freedom as the criteria by which government is to be calculated and judged. For the construction of a citizenry attached to a particular regime of self-governance has long been accorded a central significance within many different deliberations on the arts, means, and objectives of government. The promotion of forms of ethical culture, in authorities as much as in those subject to authority, may have sometimes valorized frugality, labor, obedience, and humility, but did not intend an abnegation of freedom. Indeed from the very beginnings of liberalism as a rationality of government, what was sought were the ethical techniques that would make it possible to reconcile the requirement that human beings conduct themselves simultaneously as subjects of freedom and subjects of society (Oestreich, 1982; cf. Gordon, 1991).

The 'psychologization' of technologies of the self in advanced liberal democracies needs to be understood in terms of these connections between ethics and politics. It should be seen as a constitutive feature of those social arrangements that link individuals into a social field not primarily through constraint or injunction, but through regulated acts of choice (Rose, 1992a; see this volume, Chapter 7). Within such configurations of practice, individual are presupposed who are committed to shaping a meaning for their life through the maximization of a personal 'quality of life'. Whether it be the worker in the workplace, the consumer in the marketplace, or the mother in the home, psychological vocabularies and values have enabled both political authorities and individuals to reinterpret the mundane elements of everyday life-conduct – shopping, working, cooking – as dimensions of 'life-style choice': activities *in which* people invest themselves and *through which* they both express and manifest their worth and value as selves (Rose, 1990). In these and other locales, the ethical technologies in which psychology participates, and within which psychological expertise is so deeply enmeshed, provide a means for shaping, sustaining, and managing human beings not in opposition to their personal identity but precisely in order to produce such an identity: a necessary reciprocal element of the political valorization of freedom.

Psychological expertise and liberal government

I have described in outline four of the ways in which practices bearing upon the government of persons in liberal democratic societies have been assembled in characteristically psychological forms: the administration of individuals in their singular and collective existence, the authority of authority, the relations between our notions of truth and our conduct of our selves. To discuss psychology in these terms requires us to rethink our notions of the relations between power and freedom, and the links between knowledge and subjectivity. Within our current ethical systems, the terms of each pair seem opposites, each a denial or negation of the other. But power and freedom are not antitheses (cf. Miller, 1986; Burchell, 1991; Rose, 1993). Certain modes of thinking about the exercise of power, certain techniques for endeavoring to exercise power, depend on and seek to promote freedom in those upon whom they act. Freedom is not the negation of power but one of its vital elements. We need to see how the emergence of the free, individual, and autonomous self as an idea, an ethic, and a goal to be achieved has historically been bound up with certain notions of how individuals might best be governed in accordance both with their own nature and in relation to specific objectives.

Similarly, we need to rethink the relations between knowledge and subjectivity. We tend to think of knowledge as a rationalized, sober, public domain, regulated by norms of objectivity, universality, and impartiality. This domain of knowledge would thus appear to be different in almost every respect from the private space of subjectivity, which is the play of the partial, the idiosyncratic, the experiential, the spontaneous. Of course, many anthropologists and philosophers have investigated the historicity and cultural variability of the *idea* of the individual, the person, or the self (cf. Geertz, 1979; Taylor, 1989). But I would suggest that we might understand the birth of subjectivity better when it is not merely placed in a 'cultural' context, not merely understood in terms of the history of systems of ethics and philosophy, but understood in terms of the relations of knowledge and technique. Subjectivity, that is to say, would be understood in connection with the development and transformation of modes of conceptualizing persons – vocabularies, explanatory systems, and the like – and their associated methods for acting upon persons, from the confessional to the school, from the keeping of diaries to the radio phone-in. It would be not so much a matter of holding psychology responsible for the 'social construction of subjectivity' as of seeing the ways in which the knowledges and techniques of the psychosciences have been bound up with a profound alteration in our politicoethical rationales and practices. In rendering the internality of the human being into thought, in rendering it simultaneously *visible* and *practicable,* the psychosciences have made it possible for us to dream that we can order our individual and collective existence according to a knowledge/technique that fuses truth and humanity, wisdom and practicality. It is not a question here of counterposing these dreams to a reality that they dream about, but rather of delineating the historically conditioned ways in which, in our present, experience is made amenable to thought and hence to action. Within this field, which *is* our reality, it appears that we can govern others, and govern ourselves, according to principles that are adequate to and worthy of our nature as human selves.

To conclude, let me sketch out the three principal forms of connection between psychological expertise and liberal democratic forms of government: rationality, privacy, and autonomy. First, in liberal democratic societies the exercise of power over citizens becomes legitimate to the extent that it claims a rational basis. Power is to become painstaking, calculating, and justifiable. This dependence of power upon a claim to rationality opens up a vast and auspicious territory that expertise – authority grounded in a claim to truthful knowledge and efficacious technique – can colonize: the role of psychologists in the legal and penal complex and in industry provides obvious examples.

Second, liberal democratic problematics of government depend on the creation of 'private' spaces, outside the formal scope of the authority of public powers. Yet the events within these 'private' spaces – notably the market, the organization, and the family – are construed as having vital consequences for national wealth, health, and tranquillity. The Janus face of expertise enables it to operate as a relay between government and privacy – its claims to truth and efficacy appealing, on the one hand, to governments searching for answers to their problems of regulating economic, industrial, or familial life and, on the other hand, to those in authority over these private spaces, be they industrialists or parents, attempting to manage their own private affairs efficaciously.

Third, liberal democratic problematics of government are autonomizing; they seek to govern through constructing a kind of regulated autonomy for social actors. The modern liberal self is 'obliged to be free', to construe all aspects of its life as the outcome of choices made among a number of options. Each attribute of the person is to be realized through decisions, and justified in terms of motives, needs, and aspirations of the self. The technologies of psychology gain their social power in liberal democracies because they share this ethic of competent autonomous selfhood, and because they promise to sustain, respect, and restore selfhood to citizens of such polities. They constitute technologies of individuality for the production and regulation of the individual who is 'free to choose'.

The rise of psychological expertise is made possible by the problematics of liberal democratic government, of governing through privacy, rationality, and autonomy. Hence it is appropriate to ask, in this era of social transformations, what role did the *techne* of psychology play under 'command economies' and in 'planned societies'. And it is appropriate, also, to give some consideration to the proliferation of psychological expertise in such nations in their process of transition to liberal democracies. Psychological journals are founded, especially those focusing on marriage, family, psychotherapy, and counseling. Professional organizations of psychology – of clinical psychologists, social psychologists, industrial and organizational psychologists, family and child psychologists, and so forth – are established. Institutes are established for counseling, marriage guidance, educational psychology, and much more. It appears that, as the apparatus of the party and the plan is dismantled, other forms of authority are born, other ways of shaping and guiding the choices and aspirations of these newly freed individuals in their workplaces, in their public lives, in their schools and hospitals, in their homes. Perhaps we will find that the transition to market economies and political pluralism will require, as its necessary corollary, not just the importation of the material technologies of liberal democracy but also their human technologies – the engineers of the human soul that are the other side of what we have come to term freedom.

5

Psychology as an individualizing technology

It is difficult to be precise about just how the emergence of the psychological sciences in the nineteenth century was linked up to other political and social events. This difficulty is compounded by the problem of grasping what actually does differentiate these sciences from those religious, philosophical, and medical discourses on human mental life which preceded them. In this chapter I suggest that these issues become clearer when placed in the context of 'governmentality'. The national political territories in Europe and North America in the late nineteenth and early twentieth centuries were traversed by programs for the governing of increasing areas of social and economic life in order to achieve desired objectives: security for wealth and property; continuity, efficiency, and profitability of production; public tranquillity, moral virtue, and personal responsibility. These programs were not unified by their origin in the state, or by the class allegiances of the forces that promoted them or the aims they set themselves. As we shall see later, they were as heterogeneous in conception and support as they were diverse in their strategies and mechanisms. What did characterize these programs, however, was the belief in the necessity and possibility of the management of particular aspects of social and economic existence using more or less formalized means of calculation about the relationships between means and ends: what should be done, in what ways, in order to achieve this or that desirable result. And, I suggest, it was in relation to this issue of calculability that the modern discipline of psychology was born.

Of course, my focus on these processes of calculation is hardly original. The most instructive arguments about this issue were made by Max Weber, in his analysis of capitalism as a system characterized by monetary calculations, in which capitalists make planned use of raw materials and human activities to achieve a profit on the balance sheet (Weber, 1978, esp. his comments on discipline in vol. 2, chap. 14, pt. 3). However, these programs of government did not only seek to calculate and manage financial flows, raw materials, the coordination of stages of production, and such like, but also operated on what Weber termed the 'psychophysical' apparatus of human individuals. This was motivated by the belief that achieving objectives depended on the organization of the capacities and attributes of those individuals, the ways in which they were fitted to the demands of the tasks to be undertaken, the ways in which individuals could or should be coordinated with one another in space, time, and sequence, and the means by which those lacking appropriate capacities could be identified and excluded.

These programs of government needed to forge a number of new instruments if they were to operate, the first being a new vocabulary. For the government of an enterprise or a population, a national economy or a family, a child or, indeed, oneself, it is necessary to have a way of representing the domain to be governed, its limits, characteristics, key aspects or processes, objectives, and so forth and of linking these together in some more or less systematic manner (Braudel, 1985; Tribe, 1976; Forquet, 1980; Miller and O'Leary, 1987). Although others have conceived of the languages used in regulatory practices as legitimations of the relations of power they install, this is to pose the question wrongly. Before one can seek to manage an economy, it is first necessary to conceptualize a set of processes and relations as an economy that is amenable to management. The birth of the national economy as a domain with its own characteristic laws and processes, a sphere that could be spoken about and about which knowledge could be gained, enables it to become an element in programs that seek to evaluate and increase the power of nations by governing and managing 'the economy'. Similarly, the construction of a language of the enterprise, its processes and functions, enabled the development of new forms of managerial authority over the workplace and the worker. Thus such languages do not merely legitimate power or mystify domination, they actually constitute new sectors of reality and make new aspects of existence practicable.

Psychiatry, psychology, and psychoanalysis may also be considered in this way. Two distinct but related contributions can be noted. On the one hand, these sciences provided the means for the translation of human subjectivity into a term in the new languages of government of schools, prisons, factories, the labor market, and the economy. On the other hand, they constituted the domain of subjectivity as itself a possible object for rational management, such that it became possible to conceive of desired objectives – authority, tranquillity, sanity, virtue, efficiency, and so forth – as achievable through the systematic government of subjectivity (Rose, 1985b, 1986a). For a domain to be governable, one not only needs the terms to speak about and think about it, one also needs to be able to assess its condition (cf. Latour, 1986b). That is to say, one needs intelligence or information as to what is going on in the domain one is calculating about. Information can be of various forms: written reports, drawings, pictures, numbers, charts, graphs, statistics, and so forth. It enables the features of the domain accorded pertinence – types of goods and labor, or ages of persons, their location, health, criminality – to be represented in a calculable form in the place where decisions are to be made about them: the manager's office, the war room, the case conference, the committee room of the ministry for economic affairs, or whatever. The projects for the government of social life that developed in the nineteenth century depended on and inspired the construction of moral topographies, and a statistical mapping of the population or at least its problematic sectors. The psychological sciences had a role here, in providing the devices by which human capacities and mental processes could be turned into information about which calculations could be made.

Such calculative practices are not autoeffective: they do not automatically produce functioning regulatory mechanisms and procedures. Vocabularies of calculation and accumulations of information go hand in hand with attempts to invent techniques by which the outcomes of calculative practices – in the form of decisions as to what should be done – can be translated into action upon the objects of calculation. New practices of regulation need to be constructed, and the psychological sciences made possible such procedures for the rational regulation of individuality. The management of the human factor in the institutions of modern social life could now operate in terms of a norm of truth – that is to say, in terms of a knowledge of subjectivity that had the authority of science.

The psychological sciences thus play a key role in providing the vocabulary, the information, and the regulatory techniques for the government of individuals. But we should not be misled into thinking that these features of psychiatry and psychology were invented at the behest of some all-powerful authority, or at the service of some general and more or less conscious program for control of 'deviants' (cf. Rose, 1985b). As Michel Foucault and others have argued, rather than explaining these events as responses to general and abstract social demands or functions, or as part of some inexorable process of rationalization, we need to install chance within its rightful place in history. Hitherto invisible or irrelevant aspects of conduct and behavior emerged for theoretical attention as a result of the often idiosyncratic difficulties encountered in the functioning of specific bits of social machinery. The army, the prison, the factory, the schoolroom, the family, and the community have each formed significant locations in this respect.

The figures around which concern centered often seem marginal to contemporary eyes: masturbating children and hysterical women, feebleminded schoolchildren and defective recruits to the armed forces, workers suffering fatigue or industrial accidents, unstable or shell-shocked soldiers, lying, bedwetting or naughty children. Instead of pointing to overarching strategies of the state or the professions, we need to describe the contingent and often surprising places in which these issues emerged as problems for authoritative attention, and the ways in which a variety of forces and groupings came to regard them as significant. Such strange alliances as those between socialists and nationalists over eugenics, and the equally strange oppositions, such as those between psychiatrists and psychologists over the pathologies of childhood, resist explanation in terms of a logic of class, gender, or profession. Although many of these forces have pointed to political problems and made political claims, their objectives often concern virtue as much as profit; their interests are often public good or personal happiness as much as private advancement; through their activities and inventions they actually transform the field of politics and our beliefs as to what aspects of life are administrable and by whom. Rather than bodies of professional expertise serving functions for the state, we can begin to see the way in which the very conceptions of the nature and possibilities of regulation by social authorities have been expanded and transformed.

We should not regard the role of the psychological sciences here as one of *application* of conceptual advances made in the serenity of the study or the laboratory. The impetus did not flow from an academic center to a practical periphery or from a knowledge of normality to an application to pathology. Those histories which draw us diagrams in these terms do so in order to free their subject from associations they consider disreputable or reorient it away from directions they consider unpalatable. The psychological sciences did not consolidate themselves into disciplines around the timeless project of understanding the human mind, but around contingent and historically variable problems of institutional life, the psychophysical capacities and behavioral phenomena they required and sought to produce, and the variability and vicissitudes of the human subject to which they accorded a visibility and pertinence.

Disciplining difference

In the light of these comments, let me return to the questions with which I began. What does differentiate the psychological sciences which were born in the nineteenth century from those discourses on the human soul which preceded them, and how is this difference linked up with other social and political events? This is not a question of seeking to identify some essence or founding principle lying behind all contemporary scientific concern with human mental functioning. Quite the reverse. Faced with the evident heterogeneity of the psychological sciences, their fragmented character, their lack of agreement on theory, methods, techniques, or even subject matter, and their overlaps and boundary disputes with other sciences, we need to ask ourselves how they came to be individuated as distinct disciplines. What intellectual, social, practical, or professional forces led to their partial separation from medicine, biology, philosophy, and ethics? What produced their 'disciplinization': the establishment of university departments, professorships, degree programs, laboratories, journals, training courses, professional associations, specialized employment statuses, and so forth? The authoritative histories render this question invisible through their methodological protocols and programmatic aspirations. If the necessity of the psychological sciences cannot be derived from an ontology of their object, how might we begin to understand it?

I would like to adopt a hypothesis put forward by Michel Foucault: the suggestion that all the disciplines bearing the prefix psy or psycho have their origin in what he terms a reversal of the political axis of individualization. In his book *Discipline and Punish,* Foucault writes (1977, p. 191):

> For a long time ordinary individuality – the everyday individuality of everybody – remained below the threshold of description. To be looked at, observed, described in detail, followed from day to day by an uninterrupted writing was a privilege. . . . [The disciplinary methods] reversed this relation, lowered the threshold of describable individuality and made of this description a means of control and a method of domination. . . . This turning of real lives into writing is no longer a procedure of heroization; it functions as a procedure of objectification and subjectification.

One fruitful way of thinking about the mode of functioning of the psychological sciences, and their linkages with more general social, political, and ethical transformations, might therefore be to understand them as *techniques for the disciplining of human difference:* individualizing humans through classifying then, calibrating their capacities and conducts, inscribing and recording their attributes and deficiencies, managing and utilizing their individuality and variability.

Foucault argued that the disciplines 'make' individuals by means of some rather simple technical procedures. On the parade ground, in the factory, in the school, and in the hospital, people are gathered together en masse, but by this very fact may be observed as entities both similar to and different from one another. These institutions function in certain respects like telescopes, microscopes, or other scientific instruments: they establish a regime of visibility in which the observed is distributed within a single common plane of sight. Second, these institutions operate according to a regulation of detail. These regulations, and the evaluation of conduct, manners, and so forth entailed by them, establish a grid of codeability of personal attributes. They act as norms, enabling the previously aleatory and unpredictable complexities of human conduct to be conceptually coded and cognized in terms of judgments as to conformity or deviation from such norms.

The formation of a plane of sight and a means of codeability establishes a grid of perception for registering the details of individual conduct (cf. Lynch, 1985). These have become both visible, the objects of a certain regime of visibility, and cognizable, no longer lost in the fleeting passage of space, time, movement, and voice but identifiable and notable insofar as they conform to or deviate from the network of norms that begins to spread out over the space of personal existence. Behavioral space begins to be geometrized, enabling a fixing of what was previously regarded as quintessentially unique into an ordered space of knowledge. The person is produced as a knowable individual in a process in which the properties of a disciplinary regime, its norms and values, have merged with and become attributes of persons themselves.

The individual of the psychological sciences is, to adopt a term used in another context by Michael Lynch, a 'docile' object, one that behaves (Lynch, 1985, pp. 43–4)

> in accordance with a program of normalization . . . when an object becomes observable, measurable and quantifiable, it has already become *civilized:* the disciplinary organization of civilization extends its subjection to the object in the very way it makes it knowable. The docile object provides the material template that variously supports or frustrates the operations performed upon it.

We should not, however, think of this movement from the complexities of actuality to the objectifications of knowledge as one from the concrete to the abstract. The thought object is far more concrete, far more real that its elusive 'real' referent. Persons are ephemeral, shifting, they change before one's eyes and are hard to perceive in any stable manner. Individualizing observation – be it in the institution, in the laboratory, in the clinic, in the consulting room, or psychoanalyst's office – makes the person stable through constructing a perceptual system, a way of rendering the mobile and confusing manifold of the sensible into a cognizable field. And in this perceptual process the phenomenal world is normalized – that is to say, thought in terms of its coincidences and differences from values deemed normal – in the very process of making it visible to science. Perhaps, as Foucault suggested, the psychological sciences differ from other sciences only in their low epistemological threshold. That is to say, so frequently, the norms that enable their objects to become observable, measurable, and quantifiable become part of the scientific program of perception as a consequence of having first been part of a social and institutional program of regulation – and to such programs they are destined to return.

The development of institutions and techniques that required the coordination of large numbers of persons in an economic manner and sought to eliminate certain habits, propensities, and morals and to inculcate others thus made visible the difference between those who did or did not, could or could not, would or would not learn the lessons of the institution. These institutions acted as observing and recording machines, machines for the registration of human differences. These attentions to individual differences and their consequences spread to other institutions, especially those which had to do with the efficient or rational utilization or deployment of persons. In the courtroom, in the developing system of schooling, in the apparatus concerned with pauperism and the labor market, and in the army and the factory, two sorts of problems were posed in the early years of this century, which the psychological sciences would take up. The first was a demand for some kind of human sorting house, which would assess individuals and determine to what type of regime they were best suited. The question was framed in precisely these terms in relation to delinquency, feeblemindedness, and pauperism; later, in projects for vocational guidance and selection for the armed forces. The second was the demand for advice on the ways in which individuals could best be organized and tasks best arranged so as to minimize the human problems of production or warfare – industrial accidents, fatigue, insubordination, and so forth. The consolidation of psychology into a discipline and its social destiny were tied to its capacity to produce the technical means of individualization, a new way of construing, observing, and recording human subjectivity and its vicissitudes.

Inscribing identities

Contemporaneous with the nineteenth-century transformations in the organization of asylums, prisons, hospitals, and schools, new systems were devised for documenting and recording information concerning inmates – files, records, and case histories (cf. Donnelly, 1983, chap. 7). This obligatory accumulation of the personal details and histories of large numbers of persons identifies each individual with a dossier consisting in the facts of his or her life and character accorded pertinence by the institution and its objectives. The individual here enters the field of knowledge not through any abstract leap of the philosophical imagination, but through the mundane operation of bureaucratic documentation. The sciences of individualization take off from these routine techniques of recording, utilizing them, transforming them into systematic devices for the inscription of identity, techniques that can transform the properties, capacities, energies of the human soul into material form – pictures, charts, diagrams, measurements (on inscription devices, see Latour, 1986b).

This dependence on means of visualization and techniques of inscription does not mark a fundamental difference between the psychological sciences and other sciences. Empiricist histories of science often mark the inception of the scientific conscience at a point where those who would understand the world ceased to consult the books of Aristotle and began to consult the 'book of nature'. This ingenuous fable is appealing but misleading. If scientists read the book of nature it is only because they first transformed nature into a book. Science, that is to say, not only entails techniques that render phenomena visible, so that they may form the focus of conceptualization, but also requires devices that represent the phenomena to be accounted for, which turn these phenomena into an appropriate form for analysis. Characteristically the sentence, the proposition or description in language, is not the principal mode in which phenomena are inscribed into scientific discourse in the form of evidence and data (cf. Brown and Cousins, 1980). The observation statement in linguistic form is rapidly superseded by, or at the very least accompanied by, traces of a different type: images, graphs, numbers.

Such traces produced and worked on by science have certain characteristic qualities. Latour (1986b) describes them as immutable mobiles. Whatever the original dimensions of their subjects, be they rooms full of children or chromosomes invisible to the naked eye, the traces must be neither too large nor too tiny, but of proportions that can be rapidly scanned, read, and recalled. Unlike their subjects, which are characteristically of three dimensions, and whose image is subject to variations of perspective, inscriptions are ideally of two dimensions and amenable to combination in a single visual field without variation or distortion by point of view. This enables them to be placed side by side and in various combinations, and to be integrated with materials, notes, records, and so forth from other sources. Inscriptions must render ephemeral phenomena into stable forms, which can be repeatedly examined and accumulated over time. Phenomena are frequently stuck in time and space, and inconvenient for the application of the scientist's labor; inscriptions should be easily transportable so that they can be concentrated and utilized in laboratories, clinics, and other centers of accounting, calculation, and administration.

The first techniques of visualization and inscription of human differences in the psychological sciences constituted the surface of the body as the field on which psychological pathologies were to be observed. The visual image, which in the portrait had functioned as a monument to an honored nobility, now was to become a means of grasping and calibrating the sicknesses of the soul. Doctors of the insane in the late eighteenth and nineteenth centuries, from Lavater, through Pinel and Esquirol, Bucknill and Tuke and up to the theorists of degeneracy such as Maudsley and Morel, reworked and systematized the ancient arts of physiognomy, utilizing the external proportions and characteristics of the body as the means of individualization of the pathological person. Tables and arrays of visual images, from line drawings to carefully contrived photographs, sought to establish a grammar of the body. This system of perception strove for a language in which the variations and combinations of the visualized body could be systematically mapped onto invisible mental characteristics. As Sander Gilman has pointed out, the linking of these pictorial representations with case studies in textbooks on insanity and psychopathology throughout the nineteenth century performed a vital cognitive function in linking up the theoretical and the observable, materializing the theory and idealizing the object, instructing the mind through the education of the eyes (Gilman, 1982; see also Shortland, 1985; Duden, 1991).

In phrenology, criminal anthropology, and other sciences of the soul, systems were constructed to make other aspects of the individual similarly visible and legible to the trained eye. Such systems had only a limited life-span, not because of the revelation of their internal inconsistencies or through the power of any theoretical critique, but because they failed to provide the individualizing techniques which were to be demanded of them. The coordination and regulation of large numbers of persons in the expanding apparatuses of penality, industry, education, and military life produced both a demand for new techniques and vocabularies for the managing of human difference and the conditions under which they might be invented. Capacities and attributes of the soul now became evident, which, while they affected performance at school, or predisposed to crime, or had a bearing on the success of penal regimes, or were related to efficiency in the factory or liability to breakdown in the army, were not clearly inscribed upon the surface of the body. The discipline of psychology took shape around the problem of inventing these new techniques of individualization.

Materializing the mind

The first contribution of psychology to the project of individualization was the psychological test of intelligence. The psychological test was a means of visualizing, disciplining, and inscribing difference that did not rely upon the surface of the body as the diagnostic intermediary between conduct and the psyche. The problem arose in exactly this manner in the early years of universal schooling in both England and France. A group of children suddenly became apparent who, although normal to the untrained eye, could not learn the lessons of the school. They accumulated in the lowest classes, a financial burden on the authorities, a source of concern to those who regarded the school as a vital apparatus of moralization and an affront to those who considered education to be the right of all citizens. Those seeking to discover these children first scrutinized the body as a means of diagnosis. Children would parade before the doctor who would seek to find marks of pathology: stigmata, misproportioned limbs, unbalanced nerves and muscles. But it proved difficult to align the gaze of the doctor with the requirements of the institution. Difference no longer marked itself unmistakably on the body's surface. It would have to be made legible (Rose, 1985a, pp. 90–145).

This new legibility was to be made possible by a new form of normalization: the statistical rendering of human variability through the use of the normal curve. Francis Galton, in 1883, was to produce this new technique, through the argument that the simple act of comparison of the respective amount of a particular quality or attribute possessed by two members of a group enabled the mathematization of difference. This could be represented in a simple visual form once it was assumed that all qualities in a population varied according to a regular and predictable pattern, and that the characteristics of this pattern were those established for the statistical laws of large numbers. Thus individual difference could be inscribed, and hence grasped in thought and managed in reality, by means of representing the cumulative acts of comparison in the smooth outline of the 'normal' curve. Intellectual abilities could be construed as a single dimension whose variation across the population was distributed according to precise statistical laws; the capacity of any given individual could be established in terms of his or her position within this distribution; the appropriate administrative decision could be made accordingly. The intellect had become manageable. Difference had been reduced to order, graspable through its normalization into a stable, predictable, two-dimensional trace.

The procedures for visualization and inscription of difference introduced by the discipline of psychology extend far beyond the intellect (for a related discussion, see Rose, 1990). Thus, for example, in the 1920s the child became a scientific object for psychology by means of the concept of development. The psychology of development was made possible and necessary by the clinic and the nursery school. Of course, the mind and behavior of the growing child had been an object of psychological discussion prior to the 1920s. The psychologists of development ritually acknowledged the pioneering detailed studies of the development of individual infants and children undertaken by Darwin, Preyer, Shinn, Sully, Claparedé, and Stern, as well as the observations collated under the impetus of the Child Study movement (cf. Riley, 1983). But it was argued that the problem with such investigations was their idiosyncrasy, their anecdotal quality, their lack of systematic observation, the absence of consideration of the effects of surroundings, their variable methods, their lack of comparability – in short, their lack of scientific rigor. However suggestive their reflections and observations on the ways in which the abilities of children changed over time, they did not themselves found a psychology of childhood.

But what the clinic and the nursery school made possible was precisely such a psychology. For they allowed the observation and collection of data covering numbers of children of the same ages, by skilled psychological experts, under controlled, experimental, almost laboratory conditions. They thus simultaneously allowed for standardization and for normalization – the collection of comparable information on a large number of subjects and its analysis in such a way as to construct norms. A developmental norm was a standard based upon the average abilities or performance of children of a certain age on a particular task or in a particular activity. It thus not only presented a picture of what was *normal* for children of such an age, but enabled the normality of any individual child to be assessed by comparison with this norm.

The gathering of data on children of particular ages over a certain span, and the organization of these data into age norms, enabled the norms to be arranged along an axis of time, and seen as cross sections through a continuous dimension of development. Growth and temporality could become principles of organization of a psychology of childhood. And normalization and development enabled individuals to be characterized in relation to such norms in terms of this axis of time – as 'normal', 'advanced', or 'retarded'.

The work of Arnold Gesell and his collaborators demonstrates clearly the procedures by which childhood is first made *visible,* in relation to the normalization of behavioral space within the clinic, then *inscribable* through the refinement of procedures for documenting individuality, then *assessable* through the construction of scales, charts, and observation schedules. This work involved detailed observations carried out at the Yale Psycho-Clinic, which had opened in 1911 for the assessment and treatment of children having problems at school. The children were filmed and the films transformed into still photographs. These stable, two-dimensional traces could be scrutinized and compared at leisure, in order to search out regularities. The inscriptions were normalized through the selection of 'typical instances' of different types of behavior and different ages. The photographs could be arrayed into a visual display, which summarized and condensed the multifaceted activities of the children into a form suitable to be deployed in articles, textbooks, and teaching materials.

A further transformation could then be made. Rather than the pictures having to be accompanied by captions instructing us how to read them, instruction and image were welded together into line drawings. In such drawings, the child was reduced to its essential elements: only that which was normatively pertinent was displayed. The theory had merged with the object itself – not only in a set of 'illustrations' but also in the very ways in which the psychological gaze encountered and cognized the child.

In one more transformation, the concept of development was embedded into the apparatus of a test. The combined meaning of many pictures was first condensed into a table. Behavioral items characteristic and distinctive of the various age levels were defined and organized into scales with a specification of the ages at which a given proportion of children could achieve various levels on each scale. Nonintellectual behavior was thus rendered into thought in a disciplined form, and materialized in the form of a normalizing inscription device. Norms of posture and locomotion, of vocabulary, comprehension, and conversation, of personal habits, initiative, independence, and play could now be deployed in evaluation and diagnosis. The scales provided a simultaneous means of perceiving, recording, and evaluating. They summarized all the features of the object-child deemed significant at a particular age, together with norms stating the percentage of children who could do this or that at each age. They took the form of a series of questions, through whose answers, in the form of affirmatives or negatives, the individuality of each child could be grasped in relation to the norm. These developmental scales were not simply a means of assessment. They provided a new way of thinking about childhood and a new way of seeing children, one that was rapidly spread to teachers, health workers, and parents through the scientific and popular literature. Childhood had been rendered thinkable by being made visualizable, inscribable, and assessable.

The psychological assessment materializes into a routine procedure the complex ensemble of processes whereby the individual is made inscribable.

Such procedures, be they intelligence tests, developmental scales, personality assessments, or vocational guidance interviews, seek to reproduce in the form of technical apparatus any aspects of social or personal life that have been accorded psychological pertinence (on personality, see Rose, 1986a). One no longer has to aggregate persons in large institutions and observe them for long periods of time in order to see if they manifest evaluatively significant features of behavior. Codification, mathematization, and standardization make the test a mini-laboratory for the inscription of difference. The technical device of the test, by means of which almost any psychological schema for differentiating individuals may be realized, in a stable and predictable form, in a brief period of time, in a manageable space, and at the will of the expert, is a central procedure in the practices of objectification and subjectification that are so characteristic of our modernity.

Psychological assessments use essentially the same techniques for quantifying all the qualities of the human soul. Such techniques are procedures for production of difference in an ordered form. They render human subjectivity into science as a disciplined object. As the visualization techniques discussed earlier merge the propositions of the theory as to the characteristics of the body with those of the object itself in the form of a visual display, so the test enables conducts considered evaluatively pertinent to be produced in a stable, regular, and predictable manner, to be occasioned at will. As inscription devices form a perceptual system, which enables the object to be visualized in such a way as it embodies the properties of the theory, so the test forms a realization system, which enables the properties of the theory to be embodied in the actions of the subject.

The psychological assessment produces a peculiar mode of inscription of the powers of the individual. It is a form of writing whose destiny and rationale is the dossier: a diagnosis, a profile, a score. Its results are directed toward any institutional exigency where a decision is to be made through a calculation in which the capacities or characteristics of an individual will figure. Accumulated in the file or case notes, pored over in the case conference, the courtroom, or the clinic, the inscriptions of individuality invented by the psychological sciences are thus fundamental to programs for the government of subjectivity and the management of individual difference. The procedures of visualization, individualization, and inscription that characterize the psychological sciences reverse the direction of domination between human individuals and the scientific and technical imagination. They domesticate and discipline subjectivity, transforming the intangible, changeable, apparently free-willed conduct of people into manipulable, coded, materialized, mathematized, two-dimensional traces, which may be utilized in any procedure of calculation. The human individual has become calculable and manageable.

The psychological sciences transform the ways in which inscriptions of human individuality can be produced, ordered, accumulated, and circulated.

They provide techniques of visualization and inscription of individuality which objectify their subjects by inscribing their differences from one another. Such changes in the ways in which inscriptions are produced, ordered, accumulated, and circulated do far more than make the storage and communication of information more rapid, precise, and convenient. They can be thought of in the same terms as those which have been proposed by historians of memory, writing, and printing. A number of authors have suggested that many features of modern societies previously attributed to transformations at the level of 'world view' or 'economy' can be convincingly linked to changes in the techniques of inscription (cf. Goody, 1977; Ong, 1982; Eisenstein, 1979). The invention of the alphabet, of writing, and of the printing press produced fundamental changes in the cognitive and conceptual universe of the society in question. Even such an apparently straightforward activity as making a list – only possible in a scribal culture – makes it possible to construe and analyze entities in radically new ways (Goody, 1977, pp. 52–111). We might consider other, apparently less fundamental and wide-reaching, transformations in practices of inscription in similar ways. The techniques invented by the psychological sciences provided a language in which human individuality and variability could be organized and described, and a means for representing in standard forms human mental capacities and behavioral characteristics that previously had to be described in complex and idiosyncratic language. The development of these means for visualizing and inscribing human differences transforms the intellectual universe of the scientist and the administrator. In making human subjectivity cognizable, these developments also transform the practical universe of objects and relations to which things can be done. In short, these technical developments make new areas of life thinkable and practicable.

Governing subjectivity

Drawing attention to these features of the psychological sciences – their involvement in practices of government and calculation, their role in the management of individuals – may seem to suggest that they are fundamentally coercive enterprises, seeking to control or repress subjectivity. The critical sociologies of psychiatry and psychology to which I referred earlier do tend to adopt such a position. They seek to delegitimate the contemporary psychosciences and their techniques by writing a repressive and custodial history for them. Explicitly or implicitly they promote remodeled, softer, more democratic psychologies and psychiatries, which would treat individuals with more respect and give proper weight to each person's subjectivity. Within psychology itself, radical critics have often made a similar argument (cf. Armistead, 1974; Buss, 1979). It has been suggested that the dominant paradigms in Anglo-American psychology in the twentieth century have been concerned only with the overt activity of individuals, with the extent of their adaptation to social norms, and with demonstrating how their activities could be shaped toward desirable social ends. Such a psychology, it is implied, built the rationale of technological domination of subjectivity into the very fabric of psychological knowledge. It viewed people not as agents, actively interpreting and creating their world, but merely as objects. This psychology was thus well fitted for the role of administration. It achieved social influence to the extent that it assisted in a project of social control entailing the direction, manipulation, and subjection of the individual to a constraining power.

But the practices of management of individuality I have been discussing do not work principally or exclusively by repression and domination. Such practices also, and more characteristically, seek actively to produce subjects of a certain form, to mold, shape, and organize the psyche, to fabricate individuals with particular desires and aspirations. A knowledge of subjectivity is not always locked into the mechanisms of power for the purpose of increasing coercion and constraint.

Even the psychological vocabularies and techniques that were embodied in the asylums, prisons, factories, and schools of the nineteenth century, whatever their practical effects, construed the moral order of the individual as something to be produced and regulated, not repressed. The test of intelligence was certainly linked, in the first instance, to the repressive programs of eugenics, but it was also part of a new kind of attention to the population, which sought to govern individual differences in order to maximize both individual and social efficiency. The individualizing techniques embodied in the psychologies of development and personality are not linked to a repressive project. On the contrary, they enable one to construe a form of family life, education, or production that simultaneously maximizes the capacities of individuals, their personal contentment, *and* the efficiency of the institution. The very languages, assessments, and techniques supported by the critics of 'adaptationist' psychology have made it possible to conceive of a way of managing institutional life that could forge an identification between subjective fulfillment and economic advancement, family contentment, parental commitment, and so forth. We thus cannot base a critical analysis of the power of psychology on the principle of a subjectivity to be rescued from social repression. Instead, perhaps, we should question the social relations, regulatory techniques, and ethical systems in which our personal goals have become identified with personal responsibility and subjective fulfillment.

Conclusion

I have argued that we can understand the emergence and functioning of the discipline of psychology as part of a new rationale of government entailing a particular kind of attention to human individuality. This does not, of course, amount to a claim that all concerns with human mental processes since the mid-nineteenth century have arisen in the same way. Psychology is too heterogeneous for that. The suggestion, rather, is that it was around such problems that the discipline and profession of psychology coalesced. The psychology of individualization acted as a matrix in establishing the new scientific discipline and gaining it social acceptance. This provided the conditions under which concerns that had previously fallen under the aegis of other departments of knowledge, such as philosophy, biology, medicine, or ethics, could come to construe themselves as branches of a single discipline – the positive science of man.

But such qualifications should not obscure the fundamental ways in which the psychological sciences have been implicated in the constitution of modernity. Over the past 150 years, social efficiency, well-being, happiness, and tranquillity have become construed as dependent on the production and utilization of the mental capacities and propensities of individual citizens. Regulatory systems have sought to codify, calculate, supervise, and maximize the level of functioning of individuals. They have constituted a system of individualization in terms of measurement and diagnosis rather than status and worth. The psychological sciences contribute to the language, intelligence, and techniques of these programs. They objectify their subjects by individualizing them, denoting their specificity through acts of diagnosis or of measurement. These sciences render individuals knowable through establishing relations of similarity and difference amongst them and ascertaining what each shares, or does not share, with others. They render previously ungraspable facets of human variability and potentiality thinkable. In so doing, they also make new aspects of human reality practicable. As objects of a certain regime of knowledge, human individuals become possible subjects for a certain system of power, amenable to being calculated about, having things done to them, and doing things to themselves in the name of psychological capacities and subjectivity.

6

Social psychology as a science of democracy

The social psychology that was written in the 1930s, 1940s, and 1950s makes frequent references to democracy. Gordon Allport's classic article on the historical background of modern social psychology in the first edition of the *Handbook of Social Psychology* in 1954 asserts that "the roots of modern social psychology lie in the distinctive soil of western thought and civilization," suggesting that social psychology requires the rich blend of natural and biological sciences, the tradition of free inquiry and "a philosophy and ethics of democracy" (Allport, 1954). Lewin, Lippitt, and White's famous studies of styles of leadership carried out from 1938 to 1942 at the Iowa Child Welfare Research Station sought to demonstrate the differences between experimentally created groups with a democratic atmosphere and those that were autocratic or laissez faire – the differences they found were always to the advantage of democracy (Lewin, Lippitt, and White, 1939; Lippitt, 1939, 1940). George Gallup and S. F. Rae entitled their first book on public opinion polling published in 1940 *The Pulse of Democracy* and argued that "In a democratic society the views of the majority must be regarded as the ultimate tribunal for social and political issues" (Gallup and Rae, 1940, p. 15). And in England too, J. A. C. Brown, in his much reprinted textbook *The Social Psychology of Industry,* first published in 1954, has much to say about democracy, concluding that "A genuine industrial democracy can only be based on the intelligent co-operation of primary work groups with responsibly-minded managements" (Brown, 1954, p. 301).

How should one respond to this concern with democracy? We could see it merely at the level of a homology between political and scientific cultures, perhaps unsurprising in decades marked by contestation with avowedly anti-democratic movements, forces, and ideologies. We could see it as a matter of biography, for who could wonder at American and English psychologists stressing their democratic credentials in this period, especially given the role played in the growth of the discipline by Jews fleeing Nazi Germany. Less flatteringly, we could merely see, in this rhetoric of democracy, yet another manifestation of the liberal conscience of American psychologists, their naive faith in their own institutions of government and in the progressive and enlightened contribution that could be made by rational and scientific social research to solving the problems of the day. Or we could relegate this democratic vocabulary to the periphery of the science, a matter merely of prefaces and conclusions, a kind of penumbra of social concern around a core of theoretical arguments, conceptual refinements, experiments, and proofs that answer only to the demands of scientific rigor and the test of empirical research.

I would like to suggest a different perspective, one that regards these references to democracy in the texts of social psychology as more than rhetorical flourishes. To rule citizens democratically means ruling them through their freedoms, their choices, and their solidarities rather than despite these. It means turning subjects, their motivations and interrelations, from potential sites of resistance to rule into allies of rule. It means replacing arbitrary authority with that permitting a rational justification. Social psychology as a complex of knowledges, professionals, techniques, and forms of judgment has been constitutively linked to democracy, as a way of organizing, exercising, and legitimating political power. For to rule subjects democratically it has become necessary to know them intimately.

This is not to say that psychology cannot flourish in antidemocratic politics, operate effectively in fascist societies, or be deployed in antidemocratic ways in democratic societies. It is to suggest, however, that advanced liberal democratic polities produce some characteristic problems to which the intellectual and practical technologies that comprise social psychology can promise solutions. The link between modern political culture and psychology is often supposed to lie in their shared 'individualism', and many have suggested that this accounts for the pervasive individualism of much psychological theory, including much that calls itself social psychology. It was certainly as a 'science of the individual' that psychology first found a place within the techniques of rule. In liberal democratic rationalities of government, abstract notions of the freedom of the individual are accompanied by the proliferation of rationalized practices that seek to shape, transform, and reform individuals. Thus it was not only the ethics of individualism but also the practices of individualization in the prison, the factory, the school, and the asylum that provided key conditions for the disciplinization of psychology (Rose, 1988; see this volume, Chapter 5). Psychology would find its place in all those systems where individuals were to be administered not in the light of arbitrary or willful power, but on the basis of judgments claiming objectivity, neutrality, and hence effectivity. It would provide one technology for rendering individualism operable as a set of specific programs for the regulation of existence.

The psychologies of individual differences construed the individual as a psychophysiological complex, an isolated entity to be known and predicted: a passive *object* of administration (Rose, 1985a). But other psychologies were developing that were to contribute even more to the project of governing modern societies. They sought to grasp all those phenomena and locales where persons were to be understood not in isolation but en masse – in crowds, armies, factories, families. These psychologies rendered human conduct into thought as the actions of psychological *subjects* engaged in making sense of their experience, driven by wishes, hopes, and fears, bound into dynamic relations with other persons and their social world. These social and dynamic psychologies were to offer the possibility of making democracy operable through procedures that could govern the citizen in ways consonant with the ideals of liberty, equality, and legitimate power. How this was achieved is the focus of this chapter.

Social psychology, I suggest, gains its particular role because the intellectual and practical technologies it comprises act as mechanisms that can 'translate' the principles of democracy from the domain of ethics into the sphere of scientific truths and rational expertise (cf. Callon, 1986). The social life of citizens now becomes something that can be known objectively and governed rationally. Whereas the ideals of democracy are abstract and general – respect for the individual, personal autonomy, social responsibility, the control by the people over those who govern – the power of social psychology is to enable these to be made congruent with *specific* programs for managing particular problematic areas of social life. My argument here is not based on an *interpretation* of the texts of social psychology, one that would look below their surface to discover hidden value commitments. Problems of democracy, problematizations in terms of democracy, were crucial *internal* components of social psychologists' conception of their vocation from the mid 1930s through to the mid 1960s. As Gordon Allport puts it,

> The First World War . . . followed by the spread of Communism, by the great depression of the 1930's, by the rise of Hitler, the genocide of the Jews, race riots, the Second World War and the atomic threat, stimulated all branches of social science. A special challenge fell to social psychology. The question was asked: How is it possible to preserve the values of freedom and individual rights under conditions of mounting strain and regimentation? Can science help provide an answer? This challenging question led to a burst of creative effort that added much to our understanding of the phenomena of leadership, public opinion, rumor, propaganda, prejudice, attitude change, morale, communication, decision-making, race relations and conflicts of value. (Allport, 1954, p. 2).

Social psychology was to provide a vocabulary for understanding these problems that trouble a democracy. It was to evaluate the prospects of resolving them in democratic ways. It was to provide the means for the formulation of proposals to resolve these problems that, on the one hand, were in accordance with rational scientific knowledge and, on the other hand, accorded with the democratic values of Western, liberal, pluralist, and individualist societies. And it was to contribute to the technologies that would seek to give effect to these new ways of governing.

Governing and knowing

Gordon Allport quotes Giambattista Vico at the opening of his historical review of social psychology: "government must conform to the nature of the men governed" (Vico, [1725] 1848, in Allport, 1954, p. 1). For social psychology as for political philosophy, man's social nature must be known if he is to be properly governed. 'To govern' here should be understood in its original and very broad sense, encompassing directing, ruling, controlling, mastering, guiding, leading, influencing, regulating the affairs, actions, policies or functions of others in a manner that claims or is accorded authority. As dictionary definitions indicate, governing in this context is not merely a matter of the apparatus of state: rather it is a conception of the proper scope of different authorities, the objects that they should attend to, the attributes that make authority legitimate, the techniques they should use, the ends they should seek, and so forth (Miller and Rose, 1990; Rose and Miller, 1992). From the eighteenth century onward, the tasks of authorities and rulers come to be seen *in terms of government:* not just control of a territory and its subjects, but the calculated administration of the life of the population, of each and of all. From this time forth, a multitude of programs would proliferate from politicians, philanthropists, state officials, medics, clerics, and academics identifying social ills, seeking to understand them, and proposing schemes for rectifying them in order to reduce poverty, misery, crime, madness, and vice, and to bring about efficiency, wealth, health, tranquillity, and even happiness.

It is this 'will to govern' that unifies what are otherwise a disparate series of projects – politicians trying to improve economic life by controlling the money supply, managers seeking to transform the life of the factory to increase its harmony and profitability, educators seeking to transform teaching methods in order to enhance learning, doctors seeking to educate families in order to increase health. Modern societies are traversed by this will to govern, in which it has become not just legitimate but obligatory for authorities to seek to shape and channel economic, social, and personal life to avert evil and achieve good. And the "will to govern" has constituted the self and its interrelations as a key object and resource.

We can analyze the 'governmental' role of social psychology along two dimensions: the intellectual technologies it provides and the human technologies it participates in.

Intellectual technologies

To govern requires knowledge for its possibility, its legitimacy, and its effectivity. Knowledges here should not be understood as purely contemplative phenomena, but rather as *intellectual technologies,* assemblages of ways of seeing and diagnosing, techniques of calculation and judgment. Such intellectual technologies are important for government in a number of different ways. First, to govern a domain is to exert a kind of intellectual mastery over it, to isolate a sector of reality, to identify certain characteristics and processes proper to it, to make its features notable, speakable, writable, to account for them according to certain explanatory schemes. Hence 'theories' play an essential part in reversing the relations of power between the aspiring ruler and the domain over which rule is to be exercised. Economic theories have constituted 'the economy' as a specific field with its own laws that can be distinguished from the domains of nature or politics. Sociology, in articulating a knowledge of social life, has played a key role in the constitution of 'society' as a governable entity.

Psychological theories are thus significant not simply because they have provided an abstract 'science of behavior', but rather because they have represented the self and social interaction in such a way that their properties can not only be grasped in thought but correlatively transformed in practice. Thus individual psychologies 'made up' the individual as an entity with relatively fixed levels of intelligence, personality, and the like varying across the population according to the curve traced out by a normal distribution (on 'making up' persons, see Hacking, 1986). Similarly, as we shall see, the concept of attitudes rendered intelligible the social actions of individuals in relation to a world of values – political preferences, racial prejudices, religious orientations, moral beliefs – as an ordered and measurable psychological domain. The concept of the group transformed the problematic solidarities of human crowds into specifiable processes and describable relationships. These phenomena were simultaneously *disciplined,* rendered thinkable as docile objects of a rational understanding, and *epistemologized,* located within a regime for the production, circulation, and adjudication of statements in terms of truth. The psychological constitution of the individual and the group would enable a reconciliation of the doctrines of liberty and the requirements of regulation by means of a rational knowledge and a neutral expertise.

Inscription devices

Psychology is significant for government not only as a way of construing the self and its relations but also as a set of techniques for inscribing them. Knowledge, here, takes a very material form – diagrams, graphs, tables, charts, numbers – which materializes human qualities in forms amenable to normalization and calculation (Rose, 1988; cf. Latour, 1986b). The psychological test rendered visible the invisible qualities of the human soul, distilling the multifarious attributes of the person into a single figure. These inscriptions could be compared one with another, norms could be established, evaluations carried out in relation to those norms and judgments made in the light of these. The psychological test, combining normalization, judgment, and truth, becomes a vital procedure within all those practices where the efficiency of an organization comes to be seen as dependent upon a rational utilization of the human factor. Social psychology would be crucial here too, in providing devices for inscribing the social existence of persons in ways that enable them to enter into calculations. The attitude scale, the morale survey, the sociometric diagram, the graphical representations of field theory – all these will inscribe human sociality in a form in which it could become calculable. The inscriptions of social life in a stable, mobile, comparable, and combinable form could be accumulated in government departments, personnel offices, and other *centers of calculation,* power flowing to these centers along with the inscriptions that they controlled. Decision makers could now be 'in the know' about the objects of their deliberations, could operate upon these materializations of intersubjectivity, and justify their choices in relation to 'the facts'.

Human technologies

The school, the prison, the factory, the army, the child guidance clinic – we may think of all these phenomena as *human technologies* – assemblages of diverse forces, instruments, architectural forms, and persons to achieve certain ends, be they education, punishment, production, victory, or adjustment. To the extent that these seek the calculated transformation of human conduct, they are inherently linked to those knowledges and techniques that promise to bring such a transformation about. Human technologies comprise a range of related methods for linking together, shaping, channeling, and utilizing the forces of individuals and groups in pursuit of certain objectives. Government entails the construction of such technologies for acting upon persons, enrolling initially hostile or indifferent persons and forces in the pursuit of their objectives, enlisting previously oppositional elements – sexual desire, group norms, selfishness, greed – and turning them to one's own favor, building durable associations that will allow the exercise of rule (cf. Law, 1986).

Government has depended upon the institution of a range of such technologies for rendering problematic areas of economic, social, and personal life administrable (Miller and Rose, 1990; Rose and Miller, 1992). However, government in liberal democracies takes a characteristic form. It certainly construes the health, intelligence, adjustment, and virtue of its citizens as values vital to national success. But the scope for direct political action upon citizens in their everyday lives is limited. Government cannot instruct and supervise individuals in all aspects of their conduct, nor, for liberal democratic political rationalities, should it even if it could (cf. Burchell, 1991). The 'private' domains defined by such rationalities – whether in the 'civil society' of associations and organizations, in the 'market' interactions of enterprises and entrepreneurs, in the world of work itself, or in the 'family life' of citizens – are not subject to continuous scrutiny, judgment, and normalization by political authorities. Events in these 'private' spaces are, however, governed, but indirectly – in the form of innumerable projects for channeling them in ways thought beneficial. Public authorities act on them not simply through law, through establishing an educational apparatus, a social work system, and so forth, but also by altering the financial or cultural environment within which organizations and persons make their decisions, and by encouraging them to think and calculate in certain ways. 'Private' authorities act upon them, professionals and experts not only flourishing within the apparatuses of state but also promulgating their visions of how to identify and solve problems through the sale of their expertise on the market, and through the dissemination of their messages through the industry of mass communication and popular entertainment. Liberal democracies increasingly depend on these indirect mechanisms through which the conducts, desires, and decisions of independent organizations and citizens may be aligned with the aspirations and objectives of government not through the imposition of politically determined standards, but through free choice and rational persuasion.

Psychological theories, experts, languages, and calculations have had a key role here, providing the technologies to form these alliances between citizens and their rulers, persuading, convincing, shaping the private decisions of family members, managers, owners, and entrepreneurs so that they come into alignment with public goals such as increasing profitability, efficiency, health, and adjustment. Nowhere have these promises been of more importance than in the regulation of organizational life. The organization, as a bounded and administered space, an architectural form and an enclosure for powers and hierarchies, both produces new conditions for knowing individuals and problematizes new questions concerning their management. Offices, factories, hospitals, schools, armies, prisons – all entail the calculated management of human forces in the attempt to attain objectives. First individual psychology and then social psychology were to enter this space, claiming a rational knowledge of the administration of the human factor in all fields of organizational life.

Psychological expertise weaves loose associations between the programs and objectives of authorities, the values of professionals and the personal desires of individuals. The values of the expert are grounded in truth, not politics; their credentials come from the academy and the professional organization and not from the civil service or the secret police; they promise simultaneously effectiveness for the regulator and happiness for the regulated. This is why the figure of the psychologist – in the guise of the time-and-motion man with stopwatch and chart, the assessor with tests and norms, the personnel officer with attitude surveys, the organizational consultant, the market researcher, and the counselor – has become so familiar in locales from prison to the family, from the hospital to the courtroom, from the factory to the school. But not just this – the manager, the military strategist, the teacher, the doctor, the parent, and each of us has *become* a psychologist, incorporating its vocabulary into our ways of speaking, its gaze into our ways of looking, its judgments into our calculations and decisions. Social psychology, like other modes of expertise, has provided numerous powerful yet labile devices for assembling technologies of government.

The construction of attitudes

It was in 1918 that William I. Thomas and Florian Znaniecki first defined social psychology as "the science of attitudes"; by 1935 Gordon Allport was able to describe attitudes as "probably the most distinctive and indispensable concept in contemporary American social psychology" (Thomas and Znaniecki, 1918, p. 27; Allport, 1935, p. 798). In the next quarter of a century, the notion of attitude was to become part of the basic grammar of the social psychologist's accounts of human action. As Krech, Crutchfield, and Ballachy were to put it in their introduction to social psychology in 1962, "Man's social actions – whether the actions involve religious behavior, ways of earning a living, political activity, or buying and selling goods – are directed by his attitudes" (Krech, Crutchfield, and Ballachy, 1962, p. 139). The history of the concept has been written many times, different authors seeking to use 'history' to justify, rectify, or dispense with the notion of attitude as an effective force within social psychology. Here I am not much concerned with the conceptual problems embodied in various usages of the concept: the running together of cognitive and evaluative aspects, the dispute between social and individualistic accounts, the relationship of attitudes to behavior, the debate about what features are general across social groups and what differ between individuals. I prefer to pose the issue the other way around: what has the invention of a language of attitudes enabled one to think and to do, in part because of the complex, and contradictory, domain it brings into existence for investigation and disputation. What were the problems to which the technology of attitudes would promise a solution? What was born when an attitude became an invisible psychological state rather than a visible posture of the body?

For William Thomas and Florian Znaniecki, the problem was one of social disorganization consequent upon social change: how it should be understood, how it should be responded to, what role should social science play? Their study of *The Polish Peasant in Europe and America* was financed to the tune of $50,000 by Miss Helen Culver, heiress of the founder of Hull House in Chicago with which Jane Addams was associated (my account draws upon

Madge, 1963, pp. 52–87). While the money was to investigate the problem of immigration, the study was of Polish peasants in two specific situations – those in Poland who were suffering social disorganization and reorganization; those in America who were encountering American culture. In their prefatory methodological statement the authors were explicit about their objectives: social science was to enable society to respond to social change through techniques of rational control. One the one hand, society, it appeared, had been overly reliant on crude and ineffective techniques based on legislation, acts of will that attempt to exercise power in order to force the disappearance of the undesirable or to order the appearance of the desirable. These "ordering and forbidding" techniques were no more effective than earlier magical techniques used in the attempt to control nature, for they took no account of theoretical understandings of human or social motivation. But "practical sociology," as exemplified in business, philanthropy, diplomacy, personal associations, and the like, was no better. Such practical understandings were based on inadequate and commonsense assumptions about how society operates, and were directed toward immediate practical goals. They looked at problems in isolation from the whole complexity of social facts. They had practical aims and assumed particular standards of what is desirable and undesirable, whereas the social theorist should have no prior commitment to notions of what is normal or abnormal or to particular practical considerations. The social sciences could eliminate the present situation whereby "persons of merely good will are permitted to try out on society indefinitely and irresponsibly their vague and perhaps sentimental ideas" (Thomas and Znaniecki, 1918, p. 66). As the abstract laws of physical science had been embodied in technology giving material practice a self-conscious and planful character, so Thomas and Znaniecki argued for a "social technology" that would apply the knowledge accumulated by social scientists to practical situations (p. 67).

A social technology would require a scientific knowledge of human motivation. This was what social psychology was to add to the knowledge and methods of individual psychology through studies of "conscious phenomena particular to races, nationalities, religious, political, professional groups, corresponding to special occupations and interests, provoked by special influences of a social milieu, developed by educational activities and legal measures, etc." The notion of attitude was to be the means to this knowledge, for "every manifestation of conscious life . . . can be treated as an attitude, because every one involves a tendency to action" in relation to certain objects which are "social values" (p. 27). Social psychology was to chart attitudes and their relation to values, paying particular attention to those shared among members of a social group. "Social psychology has thus to perform the part of a general science of the subjective side of social culture which we have heretofore usually ascribed to individual psychology or to 'psychology in general'" (p. 31).

An attitude was "a process of individual consciousness which determines real or possible activity of the individual in the social world" (p. 22). To a knowledge of attitudes and social values must be added a knowledge of the "definition of the situation," for it was this that would determine which attitudes were relevant to that situation and hence select one of a plurality of possible actions. The concept of attitudes thus established a connection between social values and individual activity by way of a psychological understanding: attitudes were the individual counterparts of social values, the form of connection that an individual established with a value. It was this mutual translatability of social values and individual actions by means of an intelligible space of attitudes and meanings that would be crucial for the power that attitudes would have as the organizing principle of social technologies. Attitudes could explain why, for example, the intervention of American social workers did not support the family life of Polish immigrants but undermined it (p. 48). And since "it is theoretically possible to find what social influences should be applied to certain already existing attitudes in order to produce certain new attitudes, and what attitudes should be developed with regard to certain already existing social values in order to make the individual or the group produce certain new social values, there is not a single phenomenon within the whole sphere of human life that conscious control cannot reach sooner or later" (pp. 66–7). Not that such control should treat individuals as passive objects, for "from both the moral and the hedonistic standpoints, and also from the standpoint of the level of efficiency of the individual and of the group, it is desirable to develop in the individuals the ability to control spontaneously their own activities by conscious reflection" (p. 72).

But if attitudes became intelligible in 1918, they were not yet calculable. It was in 1928 that Thurstone was to declare that "attitudes can be measured" and to inaugurate the series of inventions that would translate this subjective domain into figures that could be accumulated, compared, and calculated (Thurstone, 1928). Thurstone's little book *The Measurement of Attitudes,* published in 1929, was prefaced by some remarks by E. A. Chave on the need for religious education to be more intelligently directed to produce its desired effects in the light of a knowledge of attitudes: "Real progress," he argued, "must wait on the development of more accurate and refined objective measuring instruments" (Thurstone and Chave, 1929). It was this which Thurstone sought to achieve, "to devise a method whereby the distribution of the attitude of a group on a specified issue may be represented in the form of a frequency distribution" allowing individuals to be allocated to a position on this "attitude scale" based on the opinions that they accept or reject when questioned. The linear scale was to be a necessary part of such a method, its advantage being partly that it was consonant with the ways in which people were accustomed to ranking others on qualitative traits such as education, status, or beauty. Thurstone's scaling technique promised to transform analogous qualities into quantity, thus allowing the inscription of the extent to which any given individual, and hence a population of individuals, felt more or less favorable to prohibition, the church, capital punishment, or anything else. The power of the scale was to render into a single figure the complex intersubjective space of forces between individuals and the issues that confronted them in the social world.

By 1935 Allport could describe attitudes as the keystone in the edifice of American social psychology precisely because they formed a kind of common ground between instinct theory and environmentalism, between the dispositions of single individuals and broad patterns of culture, between psychologists and sociologists, between psychoanalysis and experimental psychology (Allport, 1935). From William James he concluded that "attitudes engender meanings upon the world." From Sigmund Freud he deduced that attitudes were endowed "with vitality, equating them with longing, hatred and love, with passion and prejudice, in short with the onrushing stream of unconscious life. . . . For the explanation of prejudice, loyalty, patriotism, crowd behavior, control by propaganda, no anemic conception of attitudes will suffice." And from Thomas and Znaniecki he discovered the importance of the social world in constituting attitudes and the role of attitudes in explaining the individual mental processes that determine the actual and potential responses of each person in the social world. An attitude, he concluded, was "a mental and neural state of readiness, organized through experience, exerting a directive or dynamic influence upon the individual's response to all objects and situations with which it is related." Whatever disputes there were between the theoreticians of attitudes and the practitioners of attitude measurement, on the complex new territory of attitudes, the dynamic forces that shape the individuals responses and actions in the social world were to become amenable to rational understanding.

This conception of attitudes was to resonate with the new political problematizations of American society in the early decades of this century, which placed great faith in the management of all areas of social life by competent, dispassionate scientific engineers, administrators, and managers (Miller and O'Leary, 1989). Progressives directed their programs of reform toward the threats that they considered were posed to democratic ideals by corrupt municipal administration, and the concentration of unaccountable power in large corporations and in the financial sector. Social scientific knowledge was to be called upon as a contributor which might help render these threats to democracy manageable, with its claims to objectivity, rationality, professionalism, and neutrality. It would thus reconcile the goals of administrative efficiency with those of democracy – authority would be exercised not through arbitrary whim or partisan interest, but on the basis of scientific exactitude. Authority was to be made reconcilable with liberal democratic ideals through scientific expertise.

The investigation of attitudes was to find its place within all these practices that sought to explore sites of social stress and develop mechanisms for understanding and regulating them. It was to provide some of the 'social intelligence' required by such bodies as the President's Research Committee on Social Trends, set up by Herbert Hoover: research was to provide "a complete impartial examination of the facts . . . to help all of us to see where social stresses are occurring and where major efforts should be undertaken to deal with them constructively. . . . The means of social control is social discovery and the wider adoption of new knowledge (Cina, 1976, p. 36, quoted in Humphreys, 1985, p. 9).

With or without the use of scales, a multitude of locales in the intersubjective world was to be mapped through the notion of attitudes: attitudes of hotel and restaurant proprietors to Chinese, attitudes of college students to Negroes, Jews, and cheating, attitudes of employees to jobs, bosses, and much else. Research on attitudes was to proceed down a number of different paths. One focused on the discrepancy between attitudes and behavior, for this problem appeared to undermine the very premises upon which attitude research had been based. A second took off from the evidence that individual attitudes often diverged from the social ideals of liberty and equality; this was to generate a prolific line of research into attitude change. A third sought to develop and deploy the technology of measurement. Thurstone's techniques for transforming quality into quantity were to be subject to a continual labor of refinement. Rensis Lickert was to simplify the matter of compiling the scales by dispensing with independent judges to decide whether items represented favorable or unfavorable attitudes; by 1939 he was using these methods in a research organization established by the U.S. Department of Agriculture. By the outbreak of World War II, the technology of attitudes was poised to fulfill the promise of a rational social technique and one in accordance with the values of democracy for which the war was fought.

Public opinion in America: Feeling the pulse of democracy

It was in the 1930s that 'the public' was discovered to have an opinion, and simultaneously that public opinion was to become construed as a vital resource for, or potential resistance to, the exercise of political rule. There was, of course, nothing new in political concern for the 'condition of the people', or in attempts to monitor it, to calculate its social consequences or even to transform it by more or less conscious political action. But in the 1930s the condition of the people was to take on a psychological hue. Earlier psychological conceptions of the mass psychology of the mob or the crowd as a potentially dangerous and irrational political force gave way to a notion of the public as an aggregate of individuals with views and wishes that could be investigated by precise techniques and communicated to government by experts (on collective psychology, see Le Bon, 1895; McDougall, 1920; and the discussions in Moscovici, 1985, and Ginneken, 1992). And the state of public opinion was to be reciprocally related to the proliferation of techniques of communication targeted on these publics, and undertaking a calculated attempt to transform them.

For writers on the psychology of collective life in the early years of this century, the crowd was qualitatively different from the individuals that made it up – less rational, more emotional, more suggestible, and hence vulnerable to the power of a strong leader. Yet, in American political culture, the psychology of the crowd was to be approached from a different perspective: not as dangerous collectivities to be tamed, but as opinions of the public that were vital to the legitimacy of a democracy. James Bryce gave public opinion a key place in *The American Commonwealth* in 1914 (Bryce, 1914; I have drawn upon Qualter, 1985). Walter Lippman and Lawrence Lowell devoted books to its discussion in the 1920s (Lowell, 1926; Lippman, 1922). Public opinion, it seemed, to be the motive force worthy of a democracy, must be truly public, not merely the uninformed assumptions of politicians or the vocal claims of unrepresentative pressure groups. But how was the real opinion of the public to be known?

Floyd Allport, in 1924, was to begin the process of disciplining the crowd, breaking it up into the behaviors and dispositions of individuals in social situations (Allport, 1924). In the decade that followed, this social psychology of individual opinion was to be brought into alignment with the technique of polling through large and systematic samples. Newspapers and magazines in the United States in the early twentieth century had gauged public opinion through "straw polls," distributing ballots to subscribers and others. The failures of this method of polling seemed to be demonstrated in the 1936 presidential election: *The Literary Digest* distributed ten million ballots and on the basis of two million returns predicted a victory for the Republican candidate, Alf Landon (Robinson, 1937). The predictions were proved wrong. But George Gallup's American Institute of Public Opinion had just been founded. Using systematic sampling techniques, it distributed 300,000 ballots among voters in all 48 states and correctly forecast the victory of Franklin D. Roosevelt. The science of opinion polling was thus validated: it was to be developed and promoted in the pages of *Public Opinion Quarterly,* which commenced publication in 1937. Floyd Allport's "Towards a science of public opinion" was its opening article, arguing against the "fallacies and blind alleys" of the old approaches to the collective mind, and proposing a view of public opinion as the arithmetical sum of individual verbalizations of favor or disfavor concerning some definite condition, person, or proposal (Allport, 1937).

Gallup, like Allport, saw public opinion not as some supraindividual social conscience, but the outgrowth of the opinions of individuals, hence calculable by polling those opinions, aggregating them, and presenting them to government and public alike. The public opinion poll, he asserted, was a crucial new instrument for taking 'the pulse of democracy', establishing the vital two-way connection between citizens and their representatives (Gallup and

Rae, 1940). In the interwar years in the United States this combination of appeals to democracy on the one hand and to science on the other proved a powerful one; polling appeared to enable the translation of democratic ideals into a rational technology of government. Not only were the opinion and consent of the policy vital legitimating forces for any system of rule, but a direct link would be opened between public opinion and governmental policy, which could supplement or even supplant the ballot box (cf. Albig, 1956).

The new way of thinking about individual will achieved through the notion of attitude was vital to this new science of democracy. No longer was the will merely an element in a speculative philosophical anthropology; it consisted in a number of specifiable and measurable attributes of individuals. Attitudes bridged the internal world of the psyche and the external world of conduct; they enabled the latter to be made intelligible and hopefully predictable in terms of the former. The advantage of the notion of attitude was that it provided a language for talking about the internal determinants of conduct and a means for thinking out how these determinants could be charted. Hence the 'attitude survey' or 'morale survey' became a key device for making the subjective world of citizens, employees, voters, and so forth inscribable and calculable. The public will could be turned into numbers and charts, which could be used in formulating arguments and strategies in the company, in the political party, in the army – indeed anywhere where individuals were to be governed 'by consent'.

A new kind of scrutiny was being applied to the mental state of the population, and this mental state was becoming linked up with the objectives of government in a new way. World War II was to prove to be a huge laboratory for the inscription, calculation, and transformation of attitudes, and of 'morale'. Morale was a powerfully mobile notion, linking up the psychiatric register, the notion of 'public opinion', the control of news and propaganda, public support for civilian and military authorities, the consequences of policy changes in army life, and much more. After America entered the war, the psychological study of morale really took off. The Society for the Psychological Study of Social Issues devoted its yearbook of 1942 to morale; the Committee for National Morale sponsored a study entitled *German Psychological Warfare,* published in 1942; the Harvard Seminar in Psychological Problems of Morale prepared worksheets that were widely circulated among government agencies; the Bureau of Intelligence of the Office of War Information made surveys for particular government agencies, as did the National Opinion Research Center, the Office of Public Opinion Research, and commercial organizations like Gallup and Fortune. Floyd and Gordon Allport studied the effects of news and rumor on civilian morale. In the light of such studies both government and private groups embarked on programs for the maintenance of morale, especially among industrial workers, conceiving military power as dependent on such phenomena as absenteeism, turnover, conflict among workers, and the adjustment of minority groups (Watson, 1942; Farago and Gittler, 1942; Katz, 1946; F. W. Allport and Lepkin, 1943; F. W. Allport, Lepkin, and Cahen, 1943; G. W. Allport and Postman, 1947; cf. the discussion in Cartwright, 1947–8). The morale of the American civilian population had become the object of knowledge; the success of administrative decisions was from now on to be construed as dependent upon information about the 'public mind'.

Perhaps it was in the U.S. Army that these inscription techniques were most closely to fulfill the dreams of Thomas and Znaniecki. The Research Branch in the Information and Education Division of the U.S. War Department, directed by Samuel A. Stouffer assisted by Lickert, was to render military decisions rationally calculable in the light of a knowledge of the attitudes of the American soldier. Its results were published in four volumes in 1949 and 1950, under the general title of Studies in Social Psychology in World War II (Stouffer et al., 1949a, 1949b, 1950a, 1950b; these studies have been much discussed, e.g., Madge, 1963). The Research Branch carried out some two hundred to three hundred large- and small-scale surveys during the war at the request of other departments. These concerned attitudes to such issues as the war, the medical services, civilians, leisure activities, army jobs, Negroes, the recruitment of women, and demobilization procedures. The work of the Research Branch appeared to show that what was crucial, from the point of view of the smooth running of an organization and the morale of individuals was indeed not the objective characteristics of the situation, but the subjective relation of the individual to his or her situation. The concept of attitude grasped this subjective relationship, the technique of scaling inscribed it. The multitudinous and diverse tastes and prejudices that mobilized the individual into this or that action could now be rendered into thought in the form of a value on each of a small number of dimensions. The linked development of the intellectual technology of attitudes and the practical technology of the scale opened this dynamic subjective world up for management; attitudes could be investigated, measured, inscribed, reported, and calculated and administrative decisions made in that light. A knowledge of attitudes made it possible to conceive of a mode of administration that would not ride roughshod over the subjective world of the individual, but incorporate it into calculations, so that persons would be managed in a way that would both conform to individual preferences and maximize organizational efficiency.

By the end of 1942, the majority of social scientists in the United States were in government service either full-time or acting as consultants on projects that blurred disciplinary boundaries, as social psychologists, anthropologists, sociologists, and psychiatrists worked together to chart and understand the social, cultural, and interpersonal determinants of the resistance of the enemy, and the conduct of the fighting forces (my discussion draws on Cartwright, 1947–8; the best general account of social scientific activity in this period is Buck, 1985). Administration would now not only calculate in terms of attitudes; it would seek to change them and to monitor the results of its attempts. This was the focus of Carl Hovland's studies, also in the Research Branch of the U.S. War Department's information division: to study the effects of the army's orientation films on the attitudes of soldiers to U.S. participation in the war (Hovland, Janis, and Kelley, 1953). This inaugurated a whole series of studies on what made a communication persuasive, and how communications could be structured to modify attitudes, that would be carried on at Yale in the postwar period.

Hovland's findings mirrored Harold Lasswell's 1948 formula of '*who* says *what* to *whom* with *what effect*'. Lasswell had been concerned about the distinction between education and propaganda since his studies in the 1920s of propaganda during World War I (Lasswell, 1926). His 1941 book on public opinion did not share Gallup's sanguine views about "the vital two-way connection between government and public opinion in a democracy." For Lasswell, public opinion could all too easily be manipulated by elites through propaganda and control of access to information (Lasswell, 1941). The management of public opinion was an aim in democracies as well as dictatorships; the growth of the technology of communication made this a real and problematic possibility. Hence it was appropriate that Lasswell should direct the Experimental Division for Study of Wartime Communication at the Library of Congress.

The study of propaganda was to be intensified during the war. German propaganda was studied by the Research Project on Totalitarian Communications at the Graduate Faculty of the New School for Social Research in New York (cf. Lerner, 1949, p. 129). And the morale of the enemy was to be investigated in many ways. Lickert complemented his work on home morale with a program conducted under the auspices of the Morale Division of the United States Strategic Bombing Survey. This systematically appraised the effects of strategic bombing on the will of countries to resist. Dorwin Cartwright, in his review of social psychology in the United States during World War II, singles out this work for particular praise. "Especially noteworthy," he concludes, "is the excellent research design involved in this project. . . . Because it is the first quantitative comparative analysis of the values and motives of people in different countries, this project will stand as a milestone in the history of social psychology" (Cartwright, 1947–8, p. 337; the research is reported in United States Strategic Bombing Survey, 1947). After the Japanese attack on Pearl Harbor in 1941, a Listening Center was established at Princeton, later developed into the Foreign Broadcast Monitoring Unit (subsequently the Foreign Broadcast Intelligence Service). Paul Lazarsfeld, Jerry Bruner, Harold Lasswell, and R. K. White developed systematic tools for the analysis of spoken and written texts, and analyzed foreign broadcasts to provide intelligence reports on military intentions and enemy morale.

This approach was to be supplemented from a rather different perspective, one that focused on the culture of the enemy and its links with the personality of the civilian and soldier. Clyde Kluckhohn, whose principal anthropological experience had been in the prewar study of the Navaho of New Mexico, and Alexander Leighton, a psychiatrist who had assisted in these studies, were appointed Joint Chiefs of the Foreign Morale Analysis Division in Washington. Leighton had previously been working at the Japanese Relocation Camp at Poston, Arizona, where he had concluded that, in times of stress, individuals appeared to behave according to belief systems: they were conditioned by the patterns and attitudes of their culture that had been deeply ingrained into them through their upbringing (Leighton, 1945; cf. Buck, 1985). In these terms, Kluckhohn and Leighton sought to understand the morale of Japanese soldiers and civilians. Basing their work on reports of interrogations of prisoners of war, captured personal and official documents, and transcripts of monitored press reports and radio broadcasts, they sought flaws in morale, which could be exploited in propaganda (Leighton, 1949). Not that they could do much to counter the belief among American policy makers that Japanese resistance to defeat was fanatical and would prevail unto death: despite their strong evidence of flagging Japanese morale by 1945, the American secretary of state for war still considered the bombing of Hiroshima and Nagasaki the only way to ensure unconditional surrender (see the discussion in Rhodes, 1987). But one outcome of this work was a remodeled psychological conception of national character which was to provide the conceptual starting point for a series of studies of 'culture and personality' in the postwar years (cf. the lucid discussion in Riley, 1983). And in the postwar period the U.S. wartime morale workers – Allport, Shils, Kluckhohn, together with Henry Murray who headed the assessment staff at the Office of Strategic Services, and Samuel Stouffer who directed the Research Branch of the Information and Education Division of the War Office – were to give an institutional form to their new interdisciplinary modes of thought: the Department of Social Relations at Harvard University, which was established in 1946 (Parsons and Shils, 1951).

In the period immediately following the end of the war, the more immediate fruits of this American labor on the public psyche were to become widely available. A host of books were published reflecting on the experience of wartime and seeking to develop from it a theory of propaganda, public opinion, the psychology of rumor and attitude change (see, e.g., Bruner, 1944; Lazarsfield, Gaudet, and Berelson, 1945; Leighton, 1945; Smith, Lasswell, and Casey, 1946; see the discussion in Lerner, 1949). The precise details of this work are not my concern. What is significant is the translation of the democratic 'public mind' into a domain accessible to knowledge via psychological expertise, accessible to calculation via psychological theory, and accessible to government via psychologically informed propaganda. This psychologization of the polity, of its attitudes, solidarities, and oppositions, its interpersonal transactions, was to establish the platform for the most significant developments of psychological expertise, and the most important transformations of rationalities of government, in the postwar years.

Governing the public mind in wartime Britain

In Britain, the psychologization of the democratic polity would take a different route; only in the postwar period would the management of public opinion take a social psychological form. Wartime Britain was not to be governed through a social psychology of attitudes, but via a social psychiatry of emotional life. Thus in the *Lancet* of August 1940 Edward Glover, then director of the London Clinic of Psychoanalysis and of the Institute for the Treatment of Delinquency, announced the birth of a 'social psychiatry': "For the first time in history a government has officially recognized that the state of public opinion is as important an index of the health of the community as a full anamnesis in the case of individual illness . . . the Ministry of Information has established that group feeling is a medico-psychological concern and that it calls for instruments of precision in diagnosis" (Glover, 1940, p. 239).

The vocabulary of this concern with the psychiatric state of the population was that of mental hygiene; its focus was the prospect of air war (this account draws upon Titmuss, 1950, and Janis, 1951). The prospect, according to Titmuss, was of "a war to be conducted by the enemy first and foremost upon the unorganized, un-uniformed and undisciplined section of the nation with the object of breaking its morale to the point of surrender" (Titmuss, 1950, p. 16). From 1924 onward it was accepted almost as a matter of course that widespread panic and neurosis would ensue among civilians as a consequence of air war. This was based on reports on the behavior of civilians under air attack in World War I. As early as 1934, Churchill brought the unprecedented dangers of some three million individuals being driven out of London by panic under the pressure of continuous air attack to the attention of the House of Commons (cf. a House of Commons Debate of 28 November 1934, quoted in Titmuss, p. 9). In 1938 leading psychiatrists from London teaching hospitals and clinics presented their report to the Ministry of Health, arguing that psychiatric casualties of air raids would exceed physical casualties by three to one: there would be some three to four million cases of acute panic, hysteria, and other neurotic conditions during the first six months of air attack. A complex organization must be set up, able to provide immediate treatment in bombed areas, twenty-four hour outpatient clinics on the outskirts of cities, special hospitals, camps and work settlements in safer areas, mobile teams of psychiatrists, and mobile child guidance clinics (Titmuss, pp. 20–1).

These anxieties proved to be unfounded (cf. Titmuss, pp. 340–51). There was no evidence of any significant increase in neurotic illness or mental disorder in Britain during the war. But the anxiety about the possible epidemic of neurosis produced its own consequences. A sustained effort was made to chart the neurotic topography of the population. Aubrey Lewis, Carlos Blacker, Philip Vernon, and C. W. Ewens were among those collating and examining statistics on the link between bombing and morale. The mental state of the population was beginning to be translated into a calculable form:

inscribed, documented, turned into statistics, graphs, charts, tables which could be pored over in political deliberations and administrative initiatives (see Lewis, 1942; Vernon, 1941; Barber, 1941; Blacker, 1946; cf. the discussion in McLaine, 1979, pp. 108ff.). Studies of reactions to air war suggested that practical organizational measures, rather than psychiatric services, were most useful in promoting the capacity to adjust. It might appear to be self-evident that good and clear information should be provided as to what to do and where to go, that food, shelter, and social services should be equitably distributed, and so forth. But the desirability of these arrangements was now not merely a matter of fairness and efficiency. They were also desirable because they furthered a psychological objective. Good government was now to have to take account of the mental health of those governed.

The issue of civilian response to air raids was but the most specific of a range of wartime programs for the government of morale, whose key administrative center in Britain was the Ministry of Information (McLaine, 1979). In the early years of the war the ministry sought to make morale governable through posters, handouts, and leaflets, usually of an exhortatory character. "Freedom Is in Peril – Defend It with All Your Might"; "Our Fighting Men Depend on You." It held public meetings and wrote letters to the press. It engaged itself in a campaign to fight the destructive effects of rumor on morale. The campaign waged by the ministry owed little to psychology (e.g., Allport and Postman, 1947). The series of posters it produced sought to produce shame, guilt, and condemnation, rather than seeing rumor as arising from specifiable psychological conditions; they vilified those who speculated about the war or discussed rumors. The unfavorable public and political response to these posters helped to discredit the very idea of explicitly and directly trying to manage morale.

Yet however clumsy its operations, the ministry realized that, in order to govern morale by whatever means, one required information. Somehow, the nebulous, ambiguous, and ill-defined notion of the morale of the population had to be inscribed and transformed into information that could be transmitted to the center, where it could be examined, evaluated, compared, and contrasted with other information and with data from earlier moments in the war, and used as the basis of calculation. How else was one to gauge what public concerns were, or how effective this or that campaign technique was? Opinion testing was to be the answer. In Britain, opinion testing had developed since World War I from the ad hoc collation of views of civil servants and MPs into a relatively methodical system. This was based on government departments and used techniques of social investigation and market research in addition to monitoring press reports and interpreting the data supplied by the networks of the Special Branch and the Secret Intelligence Services (see Middlemas, 1979, esp. chap. 12). The ministry developed and extended these techniques, utilizing three main sources to turn morale into writing and make it legible. The Home Intelligence Section gathered information at shops, cinemas, transport organizations, and Citizens Advice Bureaux. Bodies such as the Brewer's Society completed questionnaires. Data were obtained from the BBC Listeners Research Unit, from officials of political parties, from police duty room reports passed on by the Home Office. Weekly reports were provided to the section from Postal and Telegraph Censorship, whose staff numbered 10,433 in May 1941, and who sometimes scrutinized up to 200,000 letters a week. Mass Observation was able to supply regular reports on such subjects as the incidence of gas mask carrying, current rumors, the size of the shelter population, shopping habits, reactions to new films and the ministry's propaganda, and also investigated particular events such as by-elections and conditions in recently raided towns (Madge and Harrison, 1938). The Wartime Social Survey intended to interview about five thousand people per month as a representative sample of the population of Great Britain, and to obtain answers to about ten simple questions designed to test attitudes to the war situation, as well as to carry out more specific investigations of attitudes to particular issues. But these early attempts were attacked by psychologists, and criticized in press and parliament as 'snooping'. Whereas the democratic ethos in the United States could be aligned with the investigation of morale, in the United Kingdom it appeared technically difficult, intrusive, and un-British.

In 1941, the role of the ministry altered. Morale propaganda and morale investigation per se were abandoned – morale, it was said, could be left to the good sense and natural resilience of the English. The ministry concentrated instead on the management of news, and censorship. It also sought to coordinate campaigns not aimed at morale directly, but at purveying information of a practical and hygienic nature: "Coughs and Sneezes Spread Diseases," "Post Before Noon," and so forth. But this did not spell the end of the attempt to transform the petty details of the moral and psychological state of the population into information. Monitoring continued through the Home Intelligence Section and Mass Observation. Public opinion surveys were carried out by BIPO, the British Institute of Public Opinion, which, like Gallup's organization, had been established before the war. And the Wartime Social Survey was transformed. Using accredited methods of sampling, interviewing, and analysis, 101 surveys had been carried out by October 1944, involving 290,000 interviews on topics requested by other departments ranging from household methods of cooking food, through methods of getting to work to coverage and effectiveness of publicity media. In the postwar period the survey would live on in the Office of Population Census and Surveys: from now on government was to involve itself intimately with the mundane details of the everyday lives of its subjects (cf. Box and Thomas, 1944).

Keith Middlemas has argued that, over this period, the political elite superimposed a system of 'continuous contract' on the traditional cycle of general elections and party warfare: "the fine measurement of opinion and its careful management by propaganda, together with the creation of a degree of mystification about the political process" achieved public submission to the dominant political ethic, encapsulated in such fictions as 'government objectives' and 'national interest' (Middlemas, 1979, p. 369). But this opposition between manipulative opinion management and popular democracy is misleading. Rather, what one can see here is the birth of a new conception of democracy, one that articulates itself in terms of an intimate link between knowledge and citizenship, and accords a key role of scientific investigation in providing the mediations between the two. This relationship, conceived as the cornerstone of modern, enlightened democratic government, was to be a fundamental principle for much postwar social psychology of opinion and attitudes, both in Britain and the United States. On the one hand the public needs to be given information in order to discharge the duties of citizenship; on the other hand the government needs to obtain information on the needs, wants, and attitudes of the public, on its psychological as well as its physical state, in order to adjust its administrative methods and objectives and gain public cooperation. In this revised image of a democratic polity, citizenship acquires a subjective form. The citizen, no longer the passive recipient of instructions or injunctions, is to be actively engaged in the maintenance of political order and social harmony. To this process, social psychology is to make key contributions; within it it is to find powerful resources that it can mobilize for its own purposes.

Groups at work

In the 1930s, from a range of different directions, a new intersubjective entity was born – the group. The group represented a field for thought, argument, and administration that was genuinely supraindividual and yet not of the order of the crowd or the mass. The group would exist as an intermediary between the individual and the population, it would inhabit the soulless world of the organization and give it subjective meaning for the employee, it would satisfy the social needs of the atomic and fragmented self isolated with the rise of the division of labor and the decline of community, it would explain ills and could be mobilized for good, it could bring about damage in its totalitarian form and contentment and efficiency in its democratic form. In the medium of the group a new relay was found where administration in the light of psychological expertise could come into alignment with the values of democracy.

The group was first discovered in the factory. This was not the first attempt to reconcile the demands of industrial efficiency with the values of democracy. As Miller and O'Leary have shown, the writings of F. W. Taylor, widely regarded as the paradigm of manipulative psychological interventions in the workplace, were part of a wider family of political programs that sought to use scientific knowledge to advance national efficiency through making the most productive use of material and human resources (Miller and O'Leary,

1989; cf. Miller and Rose, 1995). It shared with other members of this family a belief in the improvement of the efficiency of persons through the application of expertise. It was the first of many programs to bring the internal life of the enterprise into line with the values of democracy, providing a legitimacy for management by giving it a rational basis and according it the capacity to eliminate waste and thus promote the national interest.

But by the 1920s, psychologists of industry were already distancing themselves from Taylorist technologies. Early psychological studies of industry, by and large, construed the worker as a physiological apparatus whose attributes were to be analyzed, calculated, and adjusted to the design of work – lighting, rest, pauses, bench layout, and so forth – in order to minimize fatigue and maximize efficiency. But by the 1920s it appeared to industrial psychologists like C. S. Myers that "the physiological factors involved in purely muscular fatigue are now fast becoming negligible, compared with the effects of mental and nervous fatigue, monotony, want of interest, suspicion, hostility, etc. The psychological factor must therefore be the main consideration of industry and commerce in the future" (Myers, 1927; see also Farmer, 1958, and Miller, 1986). In the interwar years these psychological factors were to become the basis of a new matrix of relations between economic regulation, management of the enterprise, and psychological expertise, which were to align the needs of industry with the satisfaction of the worker. The National Institute of Industrial Psychology, established under Myers's direction in 1921, funded by Rowntree, Cadbury, Cammell Laird, and the Trustees of Carnegie and Rockefeller, and supported by virtually the entire caucus of British psychologists, propounded a view of the nature of the productive subject, the origins of industrial discontent, and the role of industrial psychology that related the subjectivity of the worker to the demand for productivity in a new way. The worker was neither a mindless brute, nor a psychophysiological machine, but an individual with a particular psychological makeup in terms of intelligence and emotions, with fears, worries, and anxieties, whose work was hampered by boredom and worry, whose resistance to management was often founded in rational concerns, and whose productive efficiency was highly dependent upon sympathy, interest, satisfaction, and contentment.

The subjectivity of the worker was to be opened to knowledge and regulation in terms of two notions central to psychological thought and strategy in the period after World War I: individual differences and mental hygiene (cf. Rose, 1985a, chaps. 7 and 8). The worker was to be individualized in terms of his or her particular psychological makeup and idiosyncrasies and the job analyzed in terms of its demands upon the worker; human resources were to be matched to occupational demands. Vocational guidance and selection would adjust recruitment to work through a psychological calculation of suitability; movement study and analysis of periods of rest and work, the design of tasks and materials, and so on would adjust work to the psychophysiology and psychology of the worker. And the worker had a complex subjective life that needed to be understood if industry was to truly take account of the human factor. The subjectivity of the worker was the outcome of the shaping of instincts by forces and constraints in early family life; the conduct of the worker was to be explained through the relationship between the personality so formed and the industrial surroundings in which work took place. Social and industrial inefficiency and unhappiness were the outcome of failures of adjustment of the internal life of the individual to the external reality in which he or she lived and worked: in short, maladjustment. The question of industrial efficiency was, at root, one of mental hygiene. But while the mental hygienists talked in terms of the mental atmosphere of the factory and the emotional relations between workers and managers, the focus of their techniques was upon the maladjusted individual, upon efficient allocation of manpower through selection and vocational guidance, upon identifying the characteristics of the normal worker in contradistinction to the neurotic, upon treatment and prophylaxis of psychoneuroses.

Developments in the United States, identified in particular with the writings of Elton Mayo and the notion of human relations, would place the group at the heart of the psychotechnology of industry (Mayo, 1933). Mayo had initially been concerned with the body as a psychophysiological mechanism – the effects of rest pauses and the conditions of the workplace upon fatigue, accidents, and labor turnover (cf. also Viteles, 1932, and Fisher and Hanna, 1932). But the conclusions he drew from the long series of studies of the Hawthorne Works of the Western Electric Company, conducted between 1923 and 1932, were to provide a new language for interpreting the links between the conditions of work and the efficiency of production. Mayo was not directly involved in the experiments at the Hawthorne Plant, and the claims he made for them differed in significant respects from those of the researchers themselves (see the discussions in Roethlisberger and Dickson, 1939, and Whitehead, 1938). For him, what was now shown to be of significance was neither the objective exigencies and characteristics of the labor process – levels of light, hours of work, and so forth – nor even the maladjustments and psychoneuroses of individual workers, but the human relations of the enterprise: the informal group life which made it up, and the subjective interrelations which composed it. Productivity, efficiency, and contentment were now to be understood in terms of the *attitudes* of the workers to their work, their *feelings* of control over their pace of work and environment, their *sense of cohesion* within their small working group, their *beliefs* about the concern and understanding that the bosses had for their individual worth and their personal problems. This was not simply a matter of drawing attention to a complex domain of informal organization in any plant that existed in tension with its formal organization. It was also that a range of new tasks emerged to be grasped by knowledge and managed in the factory.

On the one hand, the subjective features of group relations had to be rendered into thought and made amenable to calculation. The device used here was the nondirected interview. The Hawthorne investigations, for example, involved some twenty thousand interviews whose initial purpose had been to obtain objective information. But their value, as the studies progressed, appeared to be very different: they provided, rather, a way into the emotional life of the factory, the emotional significance of particular events in the experience of the worker. Thus complaints could be analyzed into their 'manifest' content and their 'latent' content, distinguishing their 'material content' from their 'psychological form'; they were not "facts in themselves" but "symptoms or indicators of personal and social situations which needed to be explained" (Roethlisberger and Dickson, 1939, p. 269). One could thus get at the thoughts, attitudes, and sentiments among workers, foremen, supervisors, and so on that might lead to problems, dissatisfactions, and conflict. The factory was a pattern of relations between those in particular organizational positions, symbolized through social distinctions, embodying certain values and expectations and requiring delicate interpretations among all involved. Problems arose, then, not only as a result of individual maladjustment, but also where these values came into conflict with one another, or where the social equilibrium was disrupted by management seeking to impose changes without recognizing the sentiments and meanings attached to the old ways of doing things.

But, once they had been conceptualized and studied, these subjective features of work could themselves be managed to promote organizational harmony. The interviews and surveys of workers themselves had a role here, for the airing of grievances was considered to be therapeutic in itself. But more generally, the task for management was to manage the enterprise and change within it in the light of a knowledge of the values and sentiments of the work force, and to act upon these so as to make them operate for, rather than against, the interests of the firm. Personnel workers had a key role here, not only in documenting values and sentiments, but also in working out plans in the light of them, advising supervisors, and diagnosing problems of the group and individuals within it. 'Communication' became a vital instrument for realigning workers' values with management objectives, through explaining the situation, clearing up misunderstandings, and allaying fears and anxieties. Personnel workers also had a role in counseling individual employees about their difficulties to assist them in adjustment to the social organization. By such techniques management could create the internal harmony that was the condition of a happy and productive factory. Human interactions, feelings, and thoughts – the psychological relations of the individual to the group – had emerged as a new domain for management (Roethlisberger and Dickson, 1939, p. 151).

The network of interpersonal relations that constituted the group within the workplace was accorded a significance beyond work. On the one hand, it appeared to have the potential to satisfy the needs of the individual for human association, and thus have a direct relation not only for productivity and efficiency, but also for mental health. On the other hand it could have a social function. Increasing division of labor had fragmented the ties of solidarity that bound individuals together; the state was too distant from individuals to bind them into the life of society at the level of their individual consciences; hence the working group appeared a crucial mechanism for dragging otherwise isolated individuals into the "general torrent of social life" (Mayo, quoted in Miller, 1986, p. 152). This danger of fragmentation of social solidarity under the influence of the advanced division of labor and the growth of individualism was, of course, a concern of many writers in the first three decades of the twentieth century, most notably Emile Durkheim. From Mayo to the present day, the social significance of employment and unemployment, its function for individual health and social solidarity, was to be cast in this subjective form and understood in terms of the psychological value of the social experience of work.

Over this period British managers, or at least those who sought to articulate and justify a rationale for management, also came to accept the economic advantages of fostering the loyalty of employees through human relations styles of management: the emphasis on the need for the integration of the worker, on the social functions of group relations, on the effectiveness of participative managerial leadership (Child, 1969). Management came to represent itself as an expert profession, and to claim that it was not capitalist discipline but industrial efficiency that required skilled managerial control over the process of production. The legitimacy and neutrality of management were to depend not only on its basis in practical experience, but also on a scientific knowledge that would cast this experience within the framework of technical rationality. And to manage rationally, one now required a knowledge of the individual and social psychology of the worker. The language and techniques of human relations allowed management to reconcile the apparently opposing realities of the bosses' imperative of efficiency with the intelligibility of the workers' resistance to it, and to claim the capacity to transform the subjectivity of the worker from an obstacle to an ally in the quest for productivity and profit.

Groups are everywhere

The group was also discovered elsewhere. One path to the group led from Gestalt theory. Thus Muzafer Sherif found the group in his study of social norms and his findings that group norms exert a force of their own upon individual judgment (Sherif, 1936). Later he was to discover how group norms could be produced in artificially created groups, how such norms could lead to conflict between groups under conditions of competition and group frustrations, and how they would lead to hostile attitudes and actions shared among group members in relation to 'out-groups'. Here the group appeared as a powerful site of potential danger (this work is described in Sherif and Sherif, 1953).

Kurt Lewin discovered a more virtuous group in his experimental applications of 'field theory'. He produced a kind of topology of the forces that bind individuals into group life: concepts such as group atmosphere, group decision, style of leadership, group values, and we-feeling, accompanied by topographical diagrams of the fields of force traversing social space, rendered the group thinkable as an entity with a solidity and a density denied it in the individualism and behaviorism of Allport's approach to attitudes. Yet, like Allport's, Lewin's studies sought to prove that the values of democracy could be given a scientific basis: the superiority of democracy over other modes of exercising social authority could be demonstrated in an experimental setting and generalized to organizational life and to cultures as a whole. In Lewin's studies democracy was a project to be achieved, an ethical value, a scientific truth, and a psychological possibility (Lewin, 1948). And democracy could not only be proved to be advantageous, it could also be taught. Lewin and Bavelas described how one could achieve "a rapid retraining of mediocre leaders into efficient democratic leaders." Not only did this make group leaders more sensitive to the possibilities of leadership, they also "felt keenly their own greater calm and poise, after they discovered that group discipline no longer depended on their constant vigilance" (Bavelas and Lewin, 1942). In the postwar period this discovery of the group as a mechanism of training was to be institutionalized in the National Training Laboratories in Group Development that Lewin would inaugurate in 1947 (cf. Marrow, 1969). It appeared that training individuals as better leaders also made them feel like better persons, that one could fulfill oneself as a person as one made oneself a more efficient manager and a more democratic leader. This work was to be linked to the British discovery of a different kind of group. The fulcrum for this new mode of thinking about and acting upon the group in Britain was the Tavistock Clinic and its postwar offspring, the Tavistock Institute of Human Relations.

In wartime Britain the group was discovered in the form of a social psychiatry of neuroses and psychopathy. Wilfred Bion found it in the methods of group treatment he developed to resolve the problems at the Training Wing of Northfield Military Hospital in 1943 – the group relations themselves could be seen as a place where individual neuroses were most clearly manifested, and the field within which such neuroses might most effectively be understood and resolved (Bion's 1946 and 1947 articles in *Human Relations* are reprinted in Bion, 1961; cf. Ahrenfeldt, 1958; Miller and Rose, 1994). Under Bion's guidance, the observation of relations in leaderless groups was to become a powerful selection tool in the work of the War Office Selection Boards, promising a neutral and rational selection technique not vulnerable to the charges of bias and favoritism. Tom Main also discovered the group in the hospital, as a system of emotional relations embracing all staff and patients that could form the basis of a new form of therapy through the group – the therapeutic community (Main, 1946). Maxwell Jones found the group in the Mill Hill Neurosis Unit for the treatment of 'effort system' and developed his own social psychiatry, based on the notion that the patient's reactions to the hospital community mirrored his or her reactions to the community outside, hence the latter could be affected by working on the former (Jones, 1952). These group therapies were to be applied by Eric Trist and Tommy Wilson in the Transitional Communities for Social Reconnection that sought to rehabilitate ex–prisoners of war for civilian life; they would later be among those who would inaugurate the Tavistock Institute of Human Relations (cf. Miller and Rose, 1988). And they were to be applied in Maxwell Jones's therapeutic communities for the rehabilitation of chronic unemployed neurotics after the war, and later in therapeutic communities for the normalization of all those maladjusted selves who were such unpromising material for conventional psychiatry (cf. Kraupl-Taylor, 1958). Both illness and cure had merged with the relational life of the group, and it seemed that the systematic management of group life could be a powerful therapeutic tool.

Groups in democracies

In the years following the end of World War II in the United States and in Britain the idea of the group provided a means of visualization, representation, and intervention in a range of different fields. The idea of the group – its internal dynamics, its pressures on the individuals who made it up, its relations with other groups – was deployed in studies of leadership, conflict and rivalry, prejudice, conformity, and much else. Diverse techniques of group relations were developed and tried out on artificial groups in laboratory situations (cf. Cartwright and Zander, 1953). But while one could see the group everywhere, it was in the workplace that this new relational notion of institutional existence was deployed most extensively and directly in relation to the problem of democracy. It appeared that one could utilize the dynamics of group life to rethink work in a way that fused the values of democracy, productivity, and contentment. There was no antithesis between what was good for the worker and what was good for the enterprise. The worker's interest in work was more than merely that of maximizing wages and minimizing the severity of labor in terms of effort and hours. Through work, the worker obtained psychological and social benefits: fulfillment and a feeling of belonging. And as a corollary, the productive worker was one who felt satisfied and involved in work. Hence the boss's interest in the laborer should not be restricted to the technical organization of the labor process, and the establishment of effective systems of command, authority, and control. It had to encompass the happiness of the worker, the human relations of the enterprise. The new psychology of group relations drew on both the British social psychiatry of the group and the American social psychology of the group. It was not simply that certain sorts of work experience caused sickness in the worker. Nor was it simply that industrial inefficiency – low productivity, absenteeism, accidents, rapid labor turnover – stemmed from the lack of mental hygiene in the factory. It was also that certain ways of organizing work were conducive to mental health, to industrial efficiency, to social adjustment, *and* to social democracy. This psychology of the worker as a social subject was explicitly fused to a radical project for transforming the conditions of labor and the authoritative relations of the workplace in the name of ethical principles, political beliefs, industrial efficiency, and mental health.

What was at issue in these new problematizations of production was the proper relations that ought to obtain between employer and employee, between leader and led, between manager and worker, in a democracy. In the first place, the worker was, after all, a citizen inside as well as outside the workplace. But, in the second place, pleasure in work and productivity of work were fundamentally linked to the stake the worker felt he or she had in the enterprise of work: in the goals of the company, its products, and its decisions. Democratic styles of leadership, communication, and consultation within the enterprise and so forth were not only of significance because of the values they embodied, but also because of their consequences for efficiency and productivity, through the links they established between the feelings and aspirations of the worker and those of the enterprise.

This way of problematizing the workplace became significant in postwar Britain because it was linked up with a concern about production from another direction: the pressing economic requirement to increase productivity. The shortage of labor meant that there was little possibility of firms recruiting more labor, and they could not spur their labor force to greater efforts with the threat of dismissal. How, then, was productivity to be increased? To this question, the new social psychology of industry appeared to promise an answer. The writings of management theorists – Urwick and Brech on the Hawthorne experiments, Northcott on personnel management, Munroe Taylor on foremanship – were joined by more directly psychological works, in particular those of J. A. C. Brown, G. R. Taylor, and R. F. Tredgold, to proselytize for a style of management and a conception of the worker that could directly address the problem of productivity (for the managers, see Urwick and Brech, 1948; Northcott, 1945; Burns Morton, 1951, discussed in Child, 1969, and Clarke, 1950; for the psychologists, see Brown, 1954; Taylor, 1950; Tredgold, 1948, 1949; Mace, 1948; Mace and Vernon, 1953).

This new British perception of the enterprise resulted from the merging of the language and analysis of the mental hygiene movement – with its concern for the positive mental health of the worker – with the human relations picture of the organization as a network of sentiments, attitudes, meanings, and values. This was given a conceptual foundation in the social psychology of the group, especially through the importation of Lewinian notions of the advantages of democratic styles of leadership. Brown's *Social Psychology of Industry,* published in 1954 and reprinted ten times in the 1950s and 1960s, was probably the most influential example of this new British approach to the subjectivity of the worker (Brown, 1954). The worker would give his first loyalty to his primary group; if he felt the interests of the firm to clash with those of the primary group "no amount of propaganda or pleading or 'discipline' will cause him to develop feelings of loyalty towards that firm" (Taylor, 1950, p. 126). Although the formal structure of the enterprise had some importance, the informal working group was the main source of control, discipline, and values of individual workers. To manage the individual worker thus required, first and foremost, management *through* the informal group. For the worker enmeshed in his or her working group, work was not a simple financial necessity; his or her need for work and relation to it did not begin and end with the wage packet. The notion of a Protestant work ethic was misleading: work played a vital role in the mental life of the individual, a source of satisfaction as much as frustration, of self-realization as much as self-denial. Hence "we must consider the factory not so much as a place where things are produced as a place in which people spend their lives: an environment for living. . . . The work situation meets many basic human needs in a way which no other situation can approach. . . . In short, the problem with which we are faced is the *humanization of work*" (Taylor, 1950, p. 20).

Social rewards, personal contentment, and the sense of belonging were crucial here, not objective conditions themselves but the workers' *attitudes* to those conditions. "When, for example, the employees in a certain department show resentment because they have seen two supervisors talking together, it is obvious that their response cannot be fully explained in terms of the objective stimulus. It must be assumed that they have an *attitude* of suspicion towards management which makes them feel in such a situation that they are being discussed and adversely criticized" (Brown, 1954, pp. 162–3).

The attitude or morale survey was not just a technique of social research; it was a powerful new device for management. It could uncover sources of irritation among employees at an early stage to enable them to be put right; the opinions expressed could be used when policies were being formulated, changes made, and innovations such as new worker amenities planned; and the "mere fact that opinions and resentments can be expressed in this way acts as a safety valve which, even in a factory with rather poor morale, may drain away such resentment" (Brown, 1954, pp. 172–3). The internal world of the factory was becoming mapped in psychological terms, the inner feelings of workers were being transmuted into measurements about which calculations could be made; the management of the enterprise was becoming an exercise in the management of opinion.

Consent to management could be produced through the adoption of the correct techniques of circulation, presentation, and discussion of information. And it was in the financial interests of management, as well as part of its social responsibility for the creation of high productivity and low industrial conflict, to create the right attitudes, the right atmosphere, the right culture.

Attitudes could be governed in two ways: communication and leadership. The concept of communication opened up the semantic life of the enterprise for investigation and regulation in a democratic fashion. The worker did not exist in a realm of brute facts and events but in a realm of meaning, and much discontent could be traced not to the actual decisions and actions of management but to the meanings which they had for workers. Hence the "wise manager should hesitate to criticize unless he has asked himself whether he has treated his employees fairly and whether he has taken the trouble to explain the situation fully to them and allowed them to discuss it fully with him" (Brown, 1954, p. 127). Humans search for the meanings and intention behind messages, so "an order, however unpleasant, and even hated, is not resented if the situation is seen to necessitate it; whereas even a reasonable order may be bitterly resented if it is thought to give expression to a feeling of contempt" (Taylor, 1950, p. 52). In the absence of explanations, workers imparted their own meanings to such events as alterations in pay, changes in working practices, staff redeployment, and so forth, and often put the worst interpretation upon them. To cater for this need for meaning, the misunderstandings and failures of communication which were so often at the root of industrial strife needed to be minimized by establishing mechanisms for the two-way flow of information. This would make clear the intentions and objectives of management to the workers and also would make known to managers any actual or potential causes of industrial discontent. Here in the workplace, democratic government once more appeared dependent upon a knowledge of the attitudes of those governed.

Leadership had been a central issue in studies of the group since the beginning, and the argument that the superiority of democratic leadership had been demonstrated by social psychology was not reversed in the light of the experience of authoritarian leadership in its Nazi and fascist forms (cf. White and Lippitt, 1960). Hence leadership could be a further technology for the democratic management of the internal world of the factory. Low productivity, high absenteeism, accidents, and labor turnover were but rarely the product of the individual personalities of workers; social psychology had shown that the style of leadership, of the managers, supervisors, and foremen, could create or destroy the atmosphere of common purpose and high morale upon which productivity depended. Leadership had been a central concern of investigators of group dynamics; Brown drew on Lewin, Lippitt, and White's prewar experiments on leadership styles in boys' clubs to show that the effects of leadership did not depend upon the individual personalities of leaders or led: whereas autocratic leadership produced aggression or apathy and *laissez faire* leadership produced chaos, democratic leadership produced not only feelings of loyalty and belonging but also the most work of the highest quality. Democratic leadership was not only good ethics, it was also good psychology and good business. Further, as Coch and French showed in their studies of the Harwood Manufacturing Corporation in Virginia, leadership and communication were intrinsically linked: one could overcome resistance to the very necessary changes in production methods that were required in the postwar period – resistance expressed in high turnover of staff, low efficiency, restriction on output, and marked aggression towards management – simply by setting up effective methods of communication and increasing group participation (Coch and French, 1948). Democratic technologies, once again, proved not only ethically preferable but also managerially and economically optimal.

The new social psychology of the group thus promised an astonishing transformation of industry, reducing friction and increasing the technical efficiency of production in terms of the numbers of hours contributed by a given labor force as well as output per man hour.

> It is probably not too much to say that Britain could expand her national income by one-half within five years without any additional capital investment if such methods were universally adopted. . . . And none of this at the cost of driving the employee harder: on the contrary, the ordinary worker would certainly be freer and happier. . . . The paradox of industry is that if you simply aim for higher output, you get neither output nor contentment. If you work for personal happiness and development of the employee, you get this and output too. (Taylor, 1950, pp. 12–13)

The enterprise thus became a microcosm of democratic society, with the need for respect for the feelings and values of the individual, with the focus on the interdependency of leader and led, with the emphasis on communication up and down the power hierarchy and on the management of opinion. The new human technologies of subjectivity aligned the management of the enterprise with images of the enlightened government for which the war had been fought and the values of freedom, citizenship, and respect for the individual that had underpinned victory. Democracy walked hand in hand with industrial productivity and human contentment.

This concern with the relations between the mental health of the worker, the productivity of industry, and the values of democracy was also promulgated in a psychoanalytic form. The Tavistock Institute of Human Relations was established at the end of the war by those who had operated the new group techniques for selection, resettlement, and therapy in the military, and now sought to apply their experience and expertise to the problems of a peacetime community (on the Tavistock, see Miller and Rose, 1988, 1994, 1995, and forthcoming). The institute was funded by the Medical Sciences Division of the Rockefeller Foundation; from its inception, it was closely linked with Kurt Lewin's Research Center for Group Dynamics, which transferred in 1947 from the Massachusetts Institute of Technology to the University of Michigan. From 1947 onward, Lewin's group carried out a series of studies funded by the U.S. Office of Naval Research into the links between morale, teamwork, supervision, and productivity – studies inspired by the

Harwood experiments which had appeared to show that employees' resistance to change could be overcome if they were involved in its planning and execution (cf. Marrow, 1957). With the Research Center, the Tavistock founded a quarterly journal, *Human Relations,* and, like it, its defining characteristic was the application of social science expertise from a range of disciplines – psychiatry, psychology, sociology, and anthropology – to the practical problems of group life. And in the Group Relations Training Conferences run by the institute, Lewinian techniques were transposed into a characteristically British form: in Britain too the awareness of group dynamics was to become part of the personal understandings and calculations of all those in authority over others (Miller and Rose, 1994).

But it was, perhaps, in proposing new ways of organizing industry that the Tavistock approach was to be most bound up with democracy. Elliot Jaques's study of the Glacier Metal Company was probably the most influential exemplar of the Tavistock approach to industry (Jaques, 1951, pp. xiii–xiv). The psychoanalysis of the organization entailed a new role for the expert, which combined research, consultancy, and therapy. The research team investigated a problem at the behest of the organization; it worked with the organization, but it was the organization itself that had to find its way to change. Group work enabled the psychological relations of the organization to be opened for attention in a new way, and these opposing forces to be overcome by the process of 'working through'. The researcher produced psychoanalytic interpretations to the group in order to increase its insight and capacity for change. It became clear, in many cases, that the apparent problem about which the organization was concerned was only a symptom of a deeper difficulty. Working through by group discussion and interpretation built upon the constructive forces within individuals and the organization: as a result of a successful analysis, the factory as a whole could become more flexible and healthy (Jaques, 1951, pp. 308ff.).

The aim of such programs was explicitly cast in democratic form. When the Glacier Project stressed the importance of material and psychological contracts between individuals or representatives in order to reduce ambiguity, clarify boundaries, and maximize the chances of harmonious relations at work, it did so in the name of a constitutional form for the employment relation, which would enshrine the values of equity and justice in the organization of work (cf. Brown and Jaques, 1965). Other social psychologists of industry shared some of these goals. Thus Brown urged a policy of decentralization for democratically run enterprises; with the Tennessee Valley Authority and Unilever as examples, he argued that large organizations should be broken into units of a size consistent with the social psychology of the group, with powers of decision vested to a large extent in those in each plant (Brown, 1954, pp. 122–3). Taylor too sought to forge an image of work consistent with the values of democracy. The real basis of industrial democracy, he argued, is the 'grassroots democracy' established by effective communication, appropriate managerial attitudes, and so forth, "compared with which the much advertised democracy of the consultative committee, and the even more tenuous satisfactions of ownership by public boards, are of insignificant importance" (Taylor, 1950, p. 20). Hostility of unions to management merely institutionalized the dichotomy between them and hampered the progress toward a life of self-fulfillment at work based on genuine cooperation.

Such programs were to make little progress in Britain in the next two decades (for a discussion of later developments in the United Kingdom and the United States, see Miller and Rose, 1995). But in Scandinavia and Europe, a democratic organization of work on social psychological principles was to become something of a reality. Reconstruction of the internal world of the factory was linked to reconstruction of the economy and society at large. If work was a fundamental human experience and vital for the satisfaction of human needs, if the group experience of working life was a fundamental element in the satisfaction of such needs, if the large concern, whether private or state controlled, would continue to exist, if the formal ownership of a factory by state or individual did not alter the basic properties of its internal environment, if management was a specialized function and therefore hierarchies of authority were inevitable, what organization of work was consistent with both the imperatives of productivity and efficiency and the ethics of humanization, fairness, justice, and democracy? To these questions, the new social psychology of the group promised a theoretical and practical answer.

Conclusion

In 1967, Dorwin Cartwright and Alvin Zander could still preface the third edition of their comprehensive account of *Group Dynamics: Research and Theory* with an explicit reference to democracy: "A democratic society derives its strength from the effective functioning of the multitude of groups it contains. Its most valuable resources are the groups of people found in its homes, communities, schools, churches, business concerns, union halls, and various branches of government. Now, more than ever before, it is recognized that these units must perform their functions well if the larger systems are to work successfully" (Cartwright and Zander, 1967, p. vii). I have tried to suggest that these references are more than rhetoric or ideology: they are intrinsic to the problematizations within which social psychology took shape, to the ways in which it constituted its objects, problems, and vocation. I have suggested that social psychology achieved its power, in part, because of its capacity to translate between the general ideals of liberal democracy and a multitude of specific and operable programs for the calculated management of life. Social psychological expertise played an active role in the formation of new images of human subjectivity and intersubjectivity, which could bring the government of the enterprise, the organization, and the population into alignment with cultural values, social expectations, political concerns, and professional aspirations.

These new ways of construing subjectivity and intersubjectivity provided novel ways of linking 'private' realms into calculations concerning the well-being of the nation, changing the very notions of the proper sphere of politics and the activities of authorities. These images were assembled together into novel intellectual and practical technologies, and embedded in previously unthinkable programs to promote particular economic and social objectives. In equipping authority with the attributes of rationality and neutrality, social psychology and social psychologists could help address and counter criticisms of power as arbitrary or exercised in the interests of the few as opposed to the many. A social psychological knowledge of attitudes and of the group could underpin systems of rule that would maximize both social and individual welfare, endowing the citizen with social rights and government with social obligations. In appearing to align the demands of democratic government with the nature of those governed, social psychology was indeed to promise a science of democracy.

7

Governing enterprising individuals

In the summer of 1989, an advertisement began to appear regularly on the front page of one of Britain's leading serious newspapers.[1] It was for a private organization called Self-Helpline and offered a range of telephone numbers for people to ring for answers to some apparently troubling questions. There were "Emotional Problems" from "Dealing with infidelity" to "Overcoming shyness." There were "Parenthood Problems" from "My child won't sleep" to "I feel like hitting my baby." There were "Work Problems" such as "Am I in the right job" or "Becoming a supervisor." And there were "Sexual Problems" from "Impotence" to "Better orgasms." For the cost of a telephone call, callers could obtain "self-help step by step answers to dealing with your problems and improving the quality of your life." They were assured that "all messages are provided by our professionals qualified in medicine, counseling and business." And, the calls could be made anonymously, without the fear of being traced: it appeared that the problem, and its solution, was entirely a matter for one's self (Self-Helpline, 1989).

In the context of the major cultural shifts taking place in Britain and many other countries in the 1980s, the rise to political power of governments adopting the rationalities of the 'new right' and espousing the logics of neoliberalism in their reforms of macroeconomic policy, organizational culture, social welfare, and the responsibilities of citizens, this little advertisement may seem trivial. Its concerns may appear hardly germane to something as weighty as the 'enterprise culture' associated in particular with the regimes of Margaret Thatcher in the United Kingdom and Ronald Reagan in the United States, and now proving so attractive to politicians in the many former welfarist polities in Scandinavia, Australia, New Zealand, and elsewhere. Certainly most of the political and sociological analyses of these shifts focus their attention elsewhere: on postmodernization, globalization, postfordism, and the like. But the forms of political reason that, at the end of the 1980s, aspired to create an enterprise culture accorded a vital *political* value to a certain image of the human being. And I think that this image of an 'enterprising self' was so potent because it was not an idiosyncratic obsession of the right of the political spectrum. On the contrary, it resonated with basic presuppositions concerning the contemporary human being that remain widely distributed in our present, presuppositions that are embodied in the very language that we use to make persons thinkable, and in our ideals as to what people should be. These presuppositions are displayed in the little advertisement I have quoted. The self is to be a subjective being, it is to aspire to autonomy, it is to strive for personal fulfillment in its earthly life, it is to interpret its reality and destiny as a matter of individual responsibility, it is to find meaning in existence by shaping its life through acts of choice. These ways of *thinking* about humans as selves, and these ways of *judging* them, are linked to certain ways of *acting* upon such selves. The guidance of selves is no longer dependent on the authority of religion or traditional morality; it has been allocated to 'experts of subjectivity' who transfigure existential questions about the purpose of life and the meaning of suffering into technical questions of the most effective ways of managing malfunction and improving 'quality of life'.

These new practices of thinking, judging, and acting are not simply 'private' matters. They are linked to the ways in which persons figure in the political vocabulary of advanced liberal democracies – no longer as subjects with duties and obligations, but as individuals, with rights and freedoms. Specific styles of political discourse may be ephemeral, and the salvationist rhetoric of enterprise culture espoused by the British conservatism of the 1980s may fade away. But the presupposition of the autonomous, choosing, free self as the value, ideal, and objective underpinning and legitimating political activity imbues the political mentalities of the modern West, as well as those now sweeping what used to be termed Eastern Europe. How are we to evaluate it?

Notions of personhood vary greatly from culture to culture, and there are many ways of accounting for such variation, connecting personhood to religious, legal, penal, and other practices bearing upon persons, and to wider social, political, and economic arrangements. Throughout his writings, Michel Foucault suggested a number of productive ways of thinking about these issues, by linking practices bearing on the self to forms of power. Foucault's work is instructive partly because it rejects two ways in which we habitually think about power and subjectivity. We often think of power in terms of constraints that dominate, deny, and repress subjectivity. Foucault, however, analyzes power not as a negation of the vitality and capacities of individuals, but as the creation, shaping, and utilization of human beings *as* subjects. Power, that is to say, works through, and not against, subjectivity (Foucault, 1982; see Miller, 1987). Further, we think about political power largely in terms of oppositions between 'the state' and 'private life', and locate subjectivity within the latter. But Foucault conceives of power as that which traverses *all* practices – from the 'macro' to the 'micro' – through which persons are ruled, mastered, held in check, administered, steered, guided, by means of which they are led by others or have come to direct or regulate their own actions (Foucault, 1979a; Miller and Rose, 1988, 1990). To analyze the relations between 'the self' and power, then, is not a matter of lamenting the ways in which our autonomy is suppressed by the state, but of investigating the ways in which subjectivity has become an essential object, target, and resource for certain strategies, tactics, and procedures of regulation.

To consider the terms that are accorded so high a political value in our present – autonomy, fulfillment, responsibility, choice – from this perspective is certainly to question whether they mark a kind of culmination of ethical evolution. But this does not imply that we should subject these terms to a critique, for example, by claiming that the rhetoric of freedom is an ideological mask for the workings of a political system that secretly denies it. We should, rather, examine the ways in which these ideals of the self are bound up with a profoundly ambiguous set of relations between human subjects and political power. Following Foucault, I have suggested that we use the term 'government' as a portmanteau notion to encompass the multiple strategies, tactics, calculations, and reflections that have sought to 'conduct the conduct' of human beings (Foucault, 1986a; Gordon, 1986, 1987; see this volume, especially Chapters 1 and 2).

We can explore these relations along three interlinked dimensions. The first dimension, roughly 'political', Foucault termed 'governmentality', or 'mentalities of government': the complex of notions, calculations, strategies, and tactics through which diverse authorities – political, military, economic, theological, medical, and so forth – have sought to act upon the lives and conducts of each and all in order to avert evils and achieve such desirable states as health, happiness, wealth, and tranquillity (Foucault, 1979b). From at least the eighteenth century, the capacities of humans, as subjects, as citizens, as individuals, as selves, have emerged as a central target and resource for authorities. Attempts to invent and exercise different types of political rule have been intimately linked to conceptions of the nature of those who are to be ruled. The autonomous subjectivity of the modern self may seem the antithesis of political power. But Foucault's argument suggests an exploration of the ways in which this autonomization of the self is itself a central feature of contemporary governmentality.

The second dimension suggested by Foucault's writings is roughly 'institutional'. However, it entails construing institutions in a particular 'technological' way, that is to say, as '*human* technologies'. Institutions from the prison, through the asylum to the workplace, the school, and the home can be seen as practices that put in play certain assumptions and objectives concerning the human beings that inhabit them (Foucault, 1977). These are embodied in the design of institutional space, the arrangements of institutional time and activity, procedures of reward and punishment, and the operation of systems of norms and judgments. They can be thought of as 'technological' in that they seek the calculated orchestration of the activities of humans under a practical rationality directed toward certain goals. They attempt to simultaneously maximize certain capacities of individuals and constrain others in accordance with particular knowledges (medical, psychological, pedagogic) and toward particular ends (responsibility, discipline, diligence, etc.). In what ways and with what consequences are our contemporary notions of subjective autonomy and enterprise embodied within the regulatory practices of a distinctively 'modern' form of life?

The third dimension for investigation of the modern self corresponds to a roughly 'ethical' field, insofar as ethics is understood in a 'practical' way, as modes of evaluating and acting upon oneself that have obtained in different historical periods (Foucault, 1986a, 1988; see my discussion in Chapter 1 of this volume). Foucault examined these in terms of what he called 'technologies of the self', techniques "which permit individuals to effect by their own means or with the help of others a certain number of operations on their own bodies and souls, thoughts, conduct and way of being, so as to transform themselves in order to attain a certain state of happiness, purity, wisdom, perfection, or immortality" (Foucault, 1988, p. 18). Ethics are thus understood as means by which individuals come to construe, decipher, act upon themselves in relation to the true and the false, the permitted and the forbidden, the desirable and the undesirable. Along this dimension, then, we would consider the ways in which the contemporary culture of autonomous subjectivity has been embodied in our techniques for understanding and improving our selves in relation to that which is true, permitted, and desirable.

'Enterprise culture' can be understood in terms of the particular connections that it establishes between these three dimensions. For enterprise links up a seductive ethics of the self, a powerful critique of contemporary institutional and political reality, and an apparently coherent design for the radical transformation of contemporary social arrangements. In the writings of 'neoliberals' like Hayek and Friedman, the well-being of both political and social existence is to be ensured not by centralized planning and bureaucracy, but through the 'enterprising' activities and choices of autonomous entities – businesses, organizations, persons – each striving to maximize its own advantage by inventing and promoting new projects by means of individual and local calculations of strategies and tactics, costs and benefits (Hayek, 1976; Friedman, 1982; for an extended discussion, see Rose, 1993). Neoliberalism is thus more than a phenomenon at the level of political philosophy. It constitutes a mentality of government, a conception of how authorities should use their powers in order to improve national well-being, the ends they should seek, the evils they should avoid, the means they should use, and, crucially, the nature of the persons upon whom they must act.

Enterprise is such a potent language for articulating a political rationality because it can connect up these general political deliberations with the formulation of specific programs that simultaneously *problematize* organizational practices in many different social locales, and provide rationales and guidelines for transforming them. The vocabulary of enterprise thus enables a political rationality to be 'translated' into attempts to govern aspects of social, economic, and personal existence that have come to appear problematic. Enterprise here not only designates a kind of organizational form, with individual units competing with one another on the market, but more generally provides an image of a mode of activity to be encouraged in a multitude of arenas of life – the school, the university, the hospital, the GP's surgery, the factory and business organization, the family, and the apparatus of social welfare. Organizations are problematized in terms of their lack of enterprise, which epitomizes their weaknesses and their failings. Correlatively, they are to be reconstructed by promoting and utilizing the enterprising capacities of each and all, encouraging them to conduct themselves with boldness and vigor, to calculate for their own advantage, to drive themselves hard, and to accept risks in the pursuit of goals. Enterprise can thus be given a 'technological' form by experts of organizational life, engineering human relations through architecture, timetabling, supervisory systems, payment schemes, curricula, and the like to achieve economy, efficiency, excellence, and competitiveness. Contemporary regulatory practices – from those which have sought to revitalize the civil and public services by remodelling them as private or pseudoprivate agencies with budgets and targets to those which have tried to reduce long-term unemployment by turning the unemployed individual into an active job seeker – have been transformed to embody the presupposition that humans are, could be, or should be enterprising individuals, striving for fulfillment, excellence, and achievement.

Hence the vocabulary of enterprise links political rhetoric and regulatory programs to the 'self-steering' capacities of subjects themselves. Along this third dimension of political rule, enterprise forges a link between the ways we are governed by others and the ways we should govern ourselves. Enterprise here designates an array of rules for the conduct of one's everyday existence: energy, initiative, ambition, calculation, and personal responsibility. The enterprising self will make an enterprise of its life, seek to maximize its own human capital, project itself a future, and seek to shape itself in order to become that which it wishes to be. The enterprising self is thus both an active self and a calculating self, a self that calculates *about* itself and that acts *upon* itself in order to better itself. Enterprise, that is to say, designates a form of rule that is intrinsically 'ethical': good government is to be grounded in the ways in which persons govern themselves.

For many critics, this vocabulary of enterprise is obfuscating rhetoric: the apotheosis of the 'capitalist illusion' that persons are 'sovereign individuals'. Such an assessment is facile. The language of enterprise is only one way of articulating ethical presuppositions that are very widely shared; that have come to form a common ground for almost all rationalities, programs, and techniques of rule in advanced liberal democratic societies. Government in such societies is not characterized by the utopian dream of a regulative machinery that will penetrate all regions of the social body, and administer them for the common good. Rather, since at least the nineteenth century, liberal political thought has been structured by the opposition between the constitutional limits of government on the one hand, and on the other the desire to arrange things such that social and economic processes turn out for the best without the need for direct political intervention (Rose and Miller, 1992). Thus the formal limitations on the powers of 'the state' have entailed, as their corollary, the proliferation of a dispersed array of programs and mechanisms, decoupled from the direct activities of the 'public' powers, which nonetheless promise to shape events in the domains of work, the market, and the family to produce such 'public' values as wealth, efficiency, health, and well-being.

The autonomy of the self is thus not the eternal antithesis of political power, but one of the objectives and instruments of modern mentalities and strategies for the conduct of conduct. Liberal democracy, if understood as an art of government and a technology of rule, has long been bound up with the invention of techniques to *constitute* the citizens of a democratic polity with the 'personal' capacities and aspirations necessary to bear the political weight that rests on them (Rose, 1993). Governing in a liberal-democratic way means governing *through* the freedom and aspirations of subjects rather than in spite of them. The possibility of imposing 'liberal' limits on the extent and scope of 'political' rule has thus been provided by a proliferation of discourses, practices, and techniques through which self-governing capabilities can be installed in free individuals in order to bring their own ways of conducting and evaluating themselves into alignment with political objectives.

A potential, if always risky and failing, solution to the problem of the regulation of 'private' spheres produced by liberal democratic political mentalities has thus been provided through the proliferation of experts grounding their authority on knowledge and technique: medics, social workers, psychiatrists, psychologists, counselors, and advisers (Rose, 1987). Governing in a liberal democratic way depends upon the availability of such techniques that will shape, channel, organize, and direct the personal capacities and selves of individuals under the aegis of a claim to objectivity, neutrality, and technical efficacy rather than one of political partiality. Through the indirect alliances established by the apparatus of expertise, the objectives of 'liberal' government can be brought into alignment with the selves of 'democratic' citizens. And contemporary mutations in government have been made both thinkable and practicable by the multitude of technologies that have now been assembled for enjoining and emplacing the regulated freedom of autonomous selves.

Technologies of the self

Many authors have commented on the rise of a therapeutic culture of the self and sought to link this to more general political transformations. The most superficial analyses have consisted in a reprise on the familiar theme that capitalism breeds individualism, the obsession with therapy being the corollary of the illusion of atomistic self-sufficiency. More considered analyses have made similar melancholy assessments (Rieff, 1966; Lasch, 1979; MacIntyre, 1981; Bourdieu, 1984). But rather than disdaining these doomed attempts to fill the absence caused by the demise of religion, cultural solidarities, or parental authority, Foucault's approach encourages us to view therapeutics as, in certain respects, continuous with these. Therapeutics, like religion, may be analyzed as a heterogeneous array of techniques of subjectification though which human beings are urged and incited to become ethical beings, to define and regulate themselves according to a moral code, to establish precepts for conducting or judging their lives, to reject or accept moral goals.

This is not the place to trace the relations between contemporary therapeutics and earlier ethical technologies (Rose, 1990; cf. Foucault, 1985, 1986a). Let me continue to explore just one significant theme: the allocation of authority over 'the conduct of conduct' to expertise. Expertise is important in at least three respects, each distinguishing the present regime of the self from those embodied in theological injunction, moral exhortation, hygienic instruction, or appeals to utilitarian calculation. First, the grounding of authority in a claim to scientificity and objectivity establishes in a unique way the distance between systems of self-regulation and the formal organs of political power that is necessary within liberal democratic rationalities of government. Second, expertise can mobilize and be mobilized within political argument in distinctive ways, producing a new relationship between knowledge and government. Expertise comes to be accorded a particular role in the formulation of programs of government and in the technologies that seek to give them effect. Third, expertise operates through the particular relation that it has with the self-regulating capacities of subjects. For the plausibility inherent in a claim to scientificity and rationalized efficacy binds subjectivity to truth, and subjects to experts, in new and potent ways.

The advertisement with which I began operates under a significant title: "Self-Help." Although this notion has a long history, today it signifies that the regulation of personal existence is not a question of politicians seeking to impose norms of conduct through an intrusive state bureaucracy backed with legal powers. Nor is it a matter of the imposition of moral standards under a religious mandate. Self-help, today, entails an alliance between professionals claiming to provide an objective, rational answer to the question of how one should conduct a life to ensure normality, contentment, and success, and individuals seeking to shape a 'life-style', not in order to conform to social conventions but in the hope of personal happiness and an 'improved quality of life'. And the mechanism of this alliance is the market, the 'free' exchange between those with a service to sell and those who have been brought to want to buy.

Contemporary individuals are incited to live as if making a *project* of themselves: they are to *work* on their emotional world, their domestic and conjugal arrangements, their relations with employment and their techniques of sexual pleasure, to develop a 'style' of living that will maximize the worth of their existence to themselves. Evidence from the United States, Europe, and the United Kingdom suggests that the implantation of such 'identity projects', characteristic of advanced liberal democracies, is constitutively linked to the rise of a new breed of spiritual directors, 'engineers of the human soul'. Although our subjectivity might appear our most intimate sphere of experience, its contemporary intensification as a political and ethical value is intrinsically correlated with the growth of expert languages, which enable us to render our relations with our selves and others into words and into thought, and with expert techniques, which promise to allow us to transform our selves in the direction of happiness and fulfillment.

The ethics of enterprise – competitiveness, strength, vigor, boldness, outwardness, and the urge to succeed – may seem to be quite opposed to the domain of the therapeutic, which is associated with hedonism and self-centeredness. And indeed, contemporary culture is ethically pluralist: the differences that Max Weber examined between the 'styles of conduct' appropriate to different 'spheres of existence' – spiritual, economic, political, aesthetic, erotic – have not been abolished (Weber, [1915] 1948). But despite such ethical pluralism, these diverse regimes operate within a single a priori: the 'autonomization' and 'responsibilization' of the self, the instilling of a reflexive hermeneutics which will afford self-knowledge and self-mastery, and the operation of all of this under the authority of experts who claim that the self can achieve a better and happier life through the application of scientific knowledge and professional skill. The allure of expertise lies in its promise to reconcile the tensions formed across the soul of the individual who is forced concurrently to inhabit different spheres. For the new experts of the psyche promise that modes of life that appear philosophically opposed – business success and personal growth, image management and authenticity – can be brought into alignment and achieve translatability through the ethics of the autonomous, choosing, psychological self.

Freud, it will be recalled, advertised psychoanalysis thus: "You will be able to convince yourself," he wrote to an imaginary patient, "that much will be gained if we succeed in transforming hysterical misery into common unhappiness. With a mental life that has been restored to health you will be better armed against that unhappiness" (Breuer and Freud, 1895, in Freud, 1953–7, vol. 2, p. 305; the next few paragraphs draw upon evidence discussed in more detail in Rose, 1990). His successors formulate their powers rather differently.

The London Centre for Psychotherapy points out that psychotherapy takes time, yet it offers "far more fulfilling relationships and greater self expression. Family and social life, sexual partnerships and work are all likely to benefit" (London Centre for Psychotherapy, 1987). Advocates of behavioral psychotherapy hold only that "the client's 'symptoms' can be regarded as discrete psychological entities which can be removed or altered by direct means" (Mackay, 1984, p. 276). But 'therapy' is generalized to include such 'symptoms' as sexual orientation, anxiety, lack of assertiveness, and the wish to increase self-control. And 'therapy' is extended to such goals as 'greater self-awareness' which should not only 'facilitate the change process but should lead the client to reappraise his life style', 'the development of problem solving skills', and increasing 'overall perceived self-efficacy'. In the more avowedly 'humanistic' and 'alternative' therapeutic systems, from Rogers's 'client centered therapy' to Perls's 'Gestalt therapy', from Berne's 'transactional analysis' to Janov's 'primal therapy', versions of the same hope are held out: you can change, you can achieve self-mastery, you can control your own destiny, you can truly be autonomous (cf. Rose, 1990).

Become whole, become what you want, become yourself: the individual is to become, as it were, an entrepreneur of itself, seeking to maximize its own powers, its own happiness, its own quality of life, though enhancing its autonomy and then instrumentalizing its autonomous choices in the service of its life-style. The self is to style its life through acts of choice, and when it cannot conduct its life according to this norm of choice, it is to seek expert assistance. On the territory of the therapeutic, the conduct of everyday existence is recast as a series of manageable problems to be understood and resolved by technical adjustment in relation to the norm of the autonomous self aspiring to self-possession and happiness.

Therapeutics has transformed work – mental and manual – into a matter of personal fulfillment and psychical identity. The employment relationship becomes significant less for the cash reward it offers than for the subjectivity it confers or denies. An entire discourse on jobs, careers, and unemployment has taken shape, conducted in therapeutic rather than economic terms (Miller, 1986). The confident, thrusting self-images of the entrepreneur seem far from such therapeutic ethics. Yet this opposition is illusory. For therapeutics can forge alliances between the liberation of the self and the pathways to personal success, promising to break through the blockages that trap us into powerlessness and passivity, into undemanding jobs and underachievement. Hence therapeutics can appeal to both sides of the employment contract: it will make us better workers at the same time as it makes us better selves. Therapy can thus offer to free each of us from our psychic chains. We can become enterprising, take control of our careers, transform ourselves into high fliers, achieve excellence, and fulfill ourselves not *in spite of* work but *by means of* work.

Therapeutics has subjectified the mundane. Everyday life, from debt, through house purchase, childbirth, marriage, and divorce has been transformed into 'life events', remediable problems of coping and adjustment. Each is to be addressed by recognizing forces of a subjective order (fears, denials, repressions, lack of psychosocial skills) and similarly subjective consequences (neurosis, tension, stress, illness). The quotidian affairs of existence have become the occasion for introspection, confession, and management by expertise. Although this may appear to entail precisely the forms of dependency to which the spirit of enterprise is opposed, this opposition is misleading. For therapeutics, here, impels the subject to 'work' on itself and to assume responsibility for its life. It seeks to equip the self with a set of tools for the management of its affairs such that it can take control of its undertakings, define its goals, and plan to achieve its needs through its own powers.

Our contemporary regime of the self is not 'antisocial'. It construes the 'relationships' of the self with lovers, family, children, friends, and colleagues as central both to personal happiness and social efficacy. All kinds of social ills, from damaged children to ill health to disruption at work and frustration at home have come to be understood as emanating from remediable incapacities in our 'interactions' with others. Thus human interaction has been made amenable to therapeutic government, and therapists have sought to take charge of this domain of the interpersonal, knowing its laws, diagnosing its ills, prescribing the ways to conduct ourselves with others that are virtuous because they are both fulfilling and healthy. Yet, however 'social' this field may be, it can be turned to the account of the enterprising self: for in recognizing the dynamic nexus of interpersonal relations that it inhabits, selves can place these under conscious control and the self can learn the skills to shape its relations with others so that it will best fulfill its own destiny.

Freud, it has been argued, built psychoanalysis upon a tragic vision. Humans were unable to escape suffering; the duty of the living to tolerate life was denied and hampered by those who promulgated illusions that the pains of existence could be transcended to ensure happiness (Rieff, 1959; Richards, 1989). But grief, frustration, disappointment, and death pose dangers to the regime of the autonomous self, for they strike at the very images of sovereignty, self-possession, omnipotent powers, secular fulfillment, and joy through life-style to which it is welded. Hence, for the new therapeutics of finitude, suffering is not to be endured but to be reframed by expertise, to be managed as a challenge and a stimulus to the powers of the self. In transcending despair through counseling or therapy, the self can be restored to its conviction that it is master of its own existence.

Although they are heterogeneous and often originate in contexts and moralities that seem quite discrepant from the world of enterprise, each of these therapeutic systems of spiritual direction operates on an ethical terrain that can be made entirely consonant with the imperatives of the enterprising self: work on yourself, improve the quality of your life, emancipate your true self, eliminate dependency, release your potential. The healthy self is to be 'free to choose'. But in embracing such an ethic of psychological health construed in terms of autonomy we are condemned to make a project out of our own identity and we have become bound to the powers of expertise.

The presuppositions of the self

A recent British recruiting poster for the Royal Navy, on the side of a London bus, emphasized one key phrase: "choose your way of life." This is indicative of a transformation, probably most emphatic over the past couple of decades, in the types of self that are presupposed in practices for the institutional administration of individuals. For the power of the forms of knowledge and techniques that I have termed the 'expertise of subjectivity' lies in the new alliances that they make possible between the aspirations of selves and the direction of life in factory, office, airline, hospital, school, and home. The self-steering capacities of individuals are now construed as vital resources for achieving private profit, public tranquillity, and social progress, and interventions in these areas have also come to be guided by the regulatory norm of the autonomous, responsible subject, obliged to make its life meaningful through acts of choice. Attempts to manage the enterprise to ensure productivity, competitiveness, and innovation, to regulate child rearing to maximize emotional health and intellectual ability, to act upon dietary and other regimes in order to minimize disease and maximize health no longer seek to discipline, instruct, moralize, or threaten subjects into compliance. Rather, they aspire to instill and use the self-directing propensities of subjects to bring them into alliance with the aspirations of authorities.

One key site has been the workplace (Rose, 1990; Miller and Rose, 1990, 1995). A new vocabulary of the employment relation has been articulated by organizational psychologists and management consultants, in which work has been reconstrued, not as a constraint upon freedom and autonomy, but as a realm in which working subjects can express their autonomy. Workers are no longer imagined merely to endure the degradations and deprivations of labor in order to gain a wage. Nor are workers construed as social creatures seeking satisfaction of needs for solidarity and security in the group relations of the workplace. Rather, the prevailing image of the worker is of an individual in search of meaning and fulfillment, and work itself is interpreted as a site within which individuals represent, construct, and confirm their identity, an intrinsic part of a style of life.

The world of work is reconceptualized as a realm in which productivity is to be enhanced, quality assured, and innovation fostered through the active engagement of the self-fulfilling impulses of the employee, through aligning the objectives of the organization with the desires of the self. Organizations are to get the most out of their employees, not by managing group relations to maximize contentment, or by rationalizing management to ensure efficiency, but by releasing the psychological striving of individuals for autonomy and creativity and channeling them into the search of the firm for excellence and success. It now appears that individuals will ally themselves with organizational objectives to the extent that they construe them as both dependent upon and enhancing their own skills of self-realization, self-presentation, self-direction, and self-management. Expertise plays the role of relay between objectives that are economically desirable and those that are personally seductive, teaching the arts of self-realization that will enhance employees as individuals as well as workers. Economic success, career progress, and personal development intersect in this new expertise of autonomous subjectivity: work has become an essential element in the path to self-realization, and the strivings of the autonomous self have become essential allies in the path to economic success.

Reciprocally, unemployment is transformed, as the unemployed individual is characterized, in many European policies and practices as in the United States, as a 'job seeker', to be acted upon in order to maintain 'job readiness' and to avoid the risk of encouraging 'dependence' (cf. Dean, 1995). Financial support is no longer in the form of benefits provided to claimants as a matter of right, but allowances, paid to clients through a contract, which specifies that they must demonstrate their active pursuit of employment through job-search activities. As Colin Gordon puts it, "The idea of one's life as the enterprise of oneself implies that there is at least a sense in which [even when unemployed] one remains always continuously employed in (at least) that one enterprise, and that it is part of the continuous business of living to make adequate provision for the preservation, reproduction and reconstruction of one's own human capital" (Gordon, 1991, p. 44). And, as Gordon also points out, this is why the 'right to permanent retraining' in France, and similar regimes elsewhere, are able to make use of the whole panoply of techniques from the new psychological culture for assembling the capacities for self-awareness, self-presentation, and self-esteem. If the maximization of these aspects of the self is not itself thought sufficient to generate new jobs, it is argued that it will provide the key to the selection of one unemployed individual over another for those that do exist. As significantly, it is hoped that it will reduce the psychological exclusion of the unemployed person from the contemporary regime of subjectivity: unemployment is to become as much like work as possible.

A second key site for the deployment of new presuppositions concerning the self is consumption. Expertise has, once more, forged alignments between broad sociopolitical objectives, the goals of producers and the self-regulating propensities of individuals. Politicoeconomic analyses and calculations have come to stress the need for a constant expansion of consumption if economic well-being is to be maintained in the interests of the national budget, the profitability of the firm, and the maintenance of levels of employment. A complex economic terrain has taken shape, in which the success of an economy is seen as dependent on the ability of politicians, planners, and manufacturers and marketers to differentiate needs, to produce products aligned to them, and to ensure the purchasing capacity to enable acts of consumption to occur. Yet political authorities can only act indirectly upon the innumerable private acts that comprise consumption, through such measures as policies on advertising, interest, credit, and the like. It is the expertise of market research, of promotion and communication, underpinned by the knowledge and techniques of subjectivity, that provides the relays through which the aspirations of ministers, the ambitions of business, and the dreams of consumers achieve mutual translatability.

These objectives are to be achieved by instrumentalizing autonomy, and promising to promote it. Consumers are constituted as actors seeking to maximize their 'quality of life' by assembling a 'life-style' though acts of choice in a world of goods. Each commodity is imbued with a 'personal' meaning, a glow cast back upon those who purchase it, illuminating the kind of person they are, or want to come to be. Design, marketing, and image construction play a vital role in the transfiguring of goods into desires and vice versa, through the webs of meaning within which each commodity is located, the phantasies of efficacy and the dreams of pleasure that guide both product innovation and consumer demand. Through this loose assemblage of agents, calculations, techniques, images, and commodities, consumer choice can be aligned with macroeconomic objectives and business advantage: economic life can be governed and entrepreneurial aspirations realized, through the choices consumers make in their quest to fulfill themselves.

The sphere of consumption, and the mechanisms of its promotion and molding, can be extended to incorporate problems that were previously governed in other ways. Health stands as an exemplar of this transformation. Healthy bodies and hygienic homes may still be a public value and a political objective. But we no longer need state bureaucracies to enjoin healthy habits of eating, of personal hygiene, of tooth care, and the like, with compulsory inspection, subsidized incentives to eat or drink correctly, and so forth. In the new domain of consumption, individuals will *want* to be healthy, experts will instruct them on how to be so, and entrepreneurs will exploit and enhance this market for health. Health will be ensured through a combination of the market, expertise, and a regulated autonomy (Rose and Miller, 1989, 1992).

Perhaps the most striking example of the complex processes through which these new networks have been constructed and operate is the regulation of 'the family'. For some two centuries, the family has been a central ideal and mechanism for the government of the social field (Donzelot, 1979). 'Familialization' was crucial to the means whereby personal capacities and conducts could be socialized, shaped, and maximized in a manner that accorded with the moral and political principles of liberal society. From at least the mid-nineteenth century, diverse projects sought to use the human technology of the family for social ends: for eliminating illegality, curbing inebriety and restricting promiscuity, imposing restrictions upon the unbridled sensualities of adults and inculcating morality into children. These had to resolve the paradox that liberal political and philosophical thought construed the family as quintessentially private, yet simultaneously accorded it all sorts of social consequences and social duties: a concurrent 'privatization' and 'responsibilization' of the family.

Expertise resolved this basic problem at the junction of the family mechanism and the goals of liberal government. It enabled a harmonization between the promotion of the family as a locus of private aspirations and the necessity that it become a kind of 'social machine' for the production of adjusted and responsible citizens. Initially it was the malfunctioning family that was the central concern. How could one minimize the social threat such families posed without destroying them by removing their endangered members? How could one act preventively on those sectors of the population thought to harbor the seeds of social risk? Expertise was to ensure that the malfunctioning family would neither be lured into dependency by especially favorable treatment, nor forced into resistance by measures that were frankly repressive. Instead it would be instructed in health, hygiene, and normality, encouraged to see its social duties as its own concerns, and thus returned to its obligations without compromising its autonomy and its responsibility for its own members.

During our own century, attention has gradually but decisively shifted from the prevention of maladaption to the production of normality itself (Rose, 1985a). The family now will meet its social obligations through promising to meet the personal aspirations of its members, as adults construe the maximization of the physical and mental welfare of their offspring as the privileged path to their own happiness. Once such an ethic comes to govern family life, individuals can themselves evaluate and normalize their parental and conjugal conduct in terms of the images of normal mothers, fathers, parents, and families generated by expertise. Bureaucratic regulation of family life is no longer needed to ensure a harmony between social objectives and personal desires. The ethics of the active choosing self can infuse the 'private' domain that for so long appeared essentially resistant to the rationale of calculation and self-promotion. Through this new mechanism, the social field can be governed through an alliance between the powers of expertise and the wishes, hopes, and fears of the responsible, autonomous family, committed to maximizing its quality of life and to the success of family members.

The government of the self

In *The Cultural Contradictions of Capitalism,* Daniel Bell suggested that there was a fundamental opposition between the calculative relation to existence that was required within industrial capitalism and the 'cult of the self', the hedonistic culture that had apparently undercut the Protestant ethic, which provided an integrative moral foundation for society at the same time as it chimed with economic needs (Bell, 1979). The reflections in the present discussion suggest that this analysis is misleading. In the heady days of the 1960s, cults of the self promised a liberation of the individual from all mundane social constraints. But today, the therapeutic culture of the self and its experts of subjectivity offer a different freedom, a freedom to realize our potential and our dreams through reshaping the style in which we conduct our secular existence. And, correlatively, mentalities of government and technologies of regulation operate in terms of an ethic of the self that stresses not stoicism or self-denial in the service of morality and society, but the maximization of choice and self-fulfillment as the touchstone of political legitimacy and the measure of the worth of nations. For both left and right, political culture is to be reshaped to secure ways of life that are fitting for free, sovereign individuals. Neoliberalism has been a powerful contributor to this reorganization of the problematics of government, questioning, from a particular ethic of individual sovereignty, the legitimacy and the capacity of authorities to know and administer the lives of their subjects in the name of their well-being. But the neoliberal vocabulary of enterprise is only one way of articulating this more fundamental transformation in mentalities of government, in which the choices of the self have become central to the moral bases of political arguments from all parts of the political arena. Within this new political culture, the diverse and conflicting moral obligations of different spheres of life – at work, at play, in the public arena, in the family, and in sexuality – can achieve a mutual translatability, once each is articulated in terms of a self striving to make its everyday existence meaningful through the choice of its way of life.

Mentalities of government in the first half of this century operated in terms of an image of the citizen as a social being. They sought to open a kind of contract between government and citizens articulated in the language of social responsibilities and social welfare. In these forms of political thought, the individual was a locus of needs that were to be socially met if malign consequences were to be avoided, but was reciprocally to be a being to whom political, civil, and social obligations and duties were to be attached. This political rationality was translated into programs such as social insurance, child welfare, and social and mental hygiene. Pedagogic technologies from universal education to the BBC were construed as devices for forming responsible citizens. Planned and socially organized mechanisms were to weave a complex web that would bind the inhabitants of a territory into a single polity, a space of regulated freedom.

Over the past twenty five years, this rationality of government has entered a chronic crisis, manifested in the appearance of counterdiscourses from all parts of the political spectrum, left and center, as well as right. 'Welfare' is criticized as bureaucratic and inefficient, as patronizing and patriarchal, as doing nothing to tackle or redress fundamental inequalities, as a usurper of private choices and freedoms, as a violation of individual rights, and much more. These counterdiscourses are not only articulated in terms of a different vision of the respective roles of the state, the market, pluralism, civil society, and the like. They are also predicated on a different notion of the proper relations between the citizen and his or her community. Across their manifold differences, these critiques of welfare are framed in a vocabulary of individual freedom, personal choice, self-fulfillment, and initiative. Citizenship is to be active and individualistic rather than passive and dependent. The political subject is henceforth to be an individual whose citizenship is manifested through the free exercise of personal choice among a variety of options. Douglas Hurd, British home secretary in the late 1980s, may have argued that "The idea of active citizenship is a necessary complement to that of the enterprise culture" (Hurd 1989, quoted in Barnett, 1989, p. 9), while his left-wing critics argued that "there can be no such thing as an active citizen in the United Kingdom until there are *actual* citizens" (ibid., p. 11), and clamored for a written constitution and democratic rights. But all shades of political opinion now agree that citizens *should* be active and not passive, that democratic government must engage the self-activating capacities of individuals in a new governmental dispensation, that it is upon the political consciousness and commitments of individual subjects that a new politics will depend.

Such a notion of the active political subject should, I suggest, be understood in terms of its consonance with the rise of regulatory technologies that enable the subject at home and at work, in acts of consumption and pleasure, to be governed 'at a distance'. We should analyze notions like 'the active citizen' not merely as rhetoric or ideology, but in terms of the ways in which contemporary political rationalities rely upon and utilize a range of technologies that install and support the civilizing project by shaping and governing subjects and enhancing their social commitment, yet are outside the formal control of the 'public powers'. To such basic nation-forming devices as a common language, skills of literacy, and transportation networks, our century has added the mass media of communication, with their pedagogies though documentary and soap opera; opinion polls and other devices that provide reciprocal links between authorities and subjects; the regulation of life-styles through advertising, marketing, and the world of goods; *and* the experts of subjectivity. These technologies do not have their origin or principle of intelligibility in 'the state', but nonetheless have made it possible to govern in an 'advanced liberal' way, providing a plethora of indirect mechanisms that can translate the goals of political, social, and economic authorities into the choices and commitments of individuals.

Not all political subjects are embraced in the new regime of the self. Those 'on the margins', literally 'outside society', are frequently either excluded and marginalized, controlled by older, harsher ways, or maintained under the particular regimes of environmental intervention and nonintervention known as 'community care'. Yet even here, as we have seen in the case of programs for the unemployed, one may observe the utilization of very similar psychological vocabularies of diagnosis and techniques of intervention, in the logics of social skills training, in the new strategies of empowerment, in the emphasis upon the importance of self-esteem. Under very different political auspices, in the activities not only of professionals but also in antiaddiction programs, self-help organizations, and special educational programs set up by leaders of disadvantaged groups and communities, one sees the operation of a very similar image of the subject we could and should be, and the use of the same psychological and therapeutic devices for reconstructing the will on the model of enterprise, self-esteem, and self-actualization.

Neoliberalism is perhaps of enduring significance, and not merely an ephemeral or corrupt phenomenon, because it was the right, rather than the left, that succeeded in formulating a political rationality consonant with this new regime of the self. In so doing, the right established some models for ways in which political authorities could make use of human technologies through which citizens themselves can act upon themselves in order to avoid what they have come to consider undesirable and achieve what they have come to think will make them happy. In all the novel political rationalities of advanced liberalism that have emerged in the wake of the neoliberal revival of the 1980s, citizens now are no longer thought to need instruction by 'political' authorities as to how to conduct themselves and regulate their everyday existence. We can now be governed through the choices that we will ourselves make, under the guidance of cultural and cognitive authorities, in the space of regulated freedom, in our individual search for happiness, self-esteem, and self-actualization, for the fulfillment of our autonomous selves.

A critical ontology of ourselves

This investigation of the forms of self that are presupposed within modern social, economic, and political relations evokes a central question addressed by Max Weber. Wilhelm Hennis has suggested that Weber's work should be read as a sustained reflection on *Menschentum,* the history of what humans are in their nature and how human lives are conducted (Hennis, 1987). Weber thus addresses questions of enduring importance: the forms of life entailed within certain economic relations; the modes in which different religious systems and forms of religious 'association' shape and direct the practical conduct of everyday economic and vocational existence; the ways in which these and other forces, such as the modern press, mold the subjective individuality of individuals and shape their *Lebenstil* or life-styles at particular historical moments. This interpretation of Weber has been linked, by Colin Gordon, to the concerns of Michel Foucault (Gordon, 1986, 1987). In the last period of his work, Foucault returned on a number of occasions to Kant's essay of 1784 entitled *What Is Enlightenment?* (Foucault, 1986c). He argued that one of the central roles of philosophy since Kant's question was to describe the nature of our present and of ourselves in that present. To ask the question What is enlightenment? for Foucault, is to understand the importance of historical investigations into the events through which we have come to recognize ourselves and act upon ourselves as certain kinds of subjects. It is to interrogate what we have become, as subjects, in our individuality, and the nature of that present in which we are.

Such an investigation would not attempt a psychological diagnosis of the modern soul. Rather, it would seek to document the categories and explanatory schemes according to which we think ourselves, the criteria and norms we use to judge ourselves, the practices through which we act upon ourselves and one another in order to make us particular kinds of being. We would, that is to say, endeavor to describe the historical a priori of our existence as subjects. And, perhaps, we should take as a starting point the notions of subjectivity, autonomy, and freedom themselves.

In this chapter I have suggested that subjectivity is inherently linked to certain types of knowledge, that projects of autonomy are linked to the growth of expertise, and that freedom is inextricably bound up with certain ways of exercising power. But I have not intended to imply that such notions are false and should be subjected to a critique, or to recommend a nihilism that proclaims the corruption of all values. If the faithful incantation of weary political nostrums is inadequate to the task of serious analysis of the conditions and consequences of our 'age of freedom', so too is knowing sociological relativism or fashionable 'postmodern' irony. If this point requires reinforcement, it would be amply supplied by the part played by the language of freedom, individuality, and choice in the recent revolutions in Eastern Europe. Hence my aim has not been to expose or to denounce our current ethical vocabulary, but to open a space for critical reflection upon the complex practices of knowledge, power, and authority that sustain the forms of life that we have come to value, and that underpin the norms of selfhood according to which we have come to regulate our existence. To claim that values are more technical than philosophical is not to denounce all values, but it is, perhaps, to suggest the limits of philosophy as the basis for a critical understanding of ethics.

From such a perspective I have tried to indicate a general change in categories of self-understanding and techniques of self-improvement that goes beyond the political dichotomy of left and right, and which forms the ethicopolitical terrain upon which their programs must be articulated and legitimated. I have argued that the rationalities of liberal government have always been concerned with internalizing their authority in citizens through inspiring, encouraging, and inaugurating programs and techniques that will simultaneously 'autonomize' and 'responsibilize' subjects. I have suggested that, over the past century, a complex network of experts and mechanisms has taken shape, outside 'the state', but fundamentally bound up with the government of health, wealth, tranquillity, and virtue. A host of programs and technologies have come to inculcate and sustain the ethic that individuals are free to the extent that they choose a life of responsible selfhood, and have promoted the dreams of self-fulfillment through the crafting of a life-style. And I have argued that the potency of a notion of an 'enterprise culture', however short-lived its particular vocabulary might prove to be, is that it embodies a political program grounded in, and drawing upon, the new regime of the active, autonomous, choosing self.

8

Assembling ourselves

The idea of 'the self' has entered a crisis that may well be irreversible. Social theorists have written countless obituaries of the image of the human being that animated our philosophies and our ethics for so long: the universal subject, stable, unified, totalized, individualized, interiorized. For some accounts, in particular those inspired by psychoanalysis, this image was always 'imaginary': humans never existed, never could exist in this coherent and unified form – human ontology is necessarily of a creature riven at its very core. For others, this 'death of the subject' is itself a real historical event: the individual to which this image of the subject corresponded emerged only recently in a limited time-space zone, and has now been swept away by cultural change. In the place of the self, new images of subjectivity proliferate: as socially constructed, as dialogic, as inscribed upon the surface of the body, as spatialized, decentered, multiple, nomadic, created in episodic recognition-seeking practices of self-display in particular times and places.

The hypotheses that I have put forward in this book should certainly be located within this problem space. However, I have also tried to address myself to a reciprocal set of problems concerning 'the self': the peculiar fact that at the very moment when this image of the human being is pronounced passé by social theorists, regulatory practices seek to govern individuals in a way more tied to their 'selfhood' than ever before, and the ideas of identity and its cognates have acquired an increased salience in so many of the practices in which human beings engage. In political life, in work, in conjugal and domestic arrangements, in consumption, marketing, and advertising, in television and cinema, in the legal complex and the practices of the police, and in the apparatuses of medicine and health, human beings are addressed, represented, and acted upon *as if they were selves* of a particular type: suffused with an individualized subjectivity, motivated by anxieties and aspirations concerning their self-fulfillment, committed to finding their true identities and maximizing their authentic expression in their life-styles. The images of freedom and autonomy that inspire our political thinking equally operate in terms of an image of each human being as the unified psychological focus of his or her biography, as the locus of legitimate rights and demands, as an actor seeking to 'enterprise' his or her life and self through acts of choice. And it appears, from the popularity of the problematics of psy in the media, in demands for therapy and the swarming of counselors, as if human beings, at least in certain places and among certain sectors, have come to recognize themselves in such images and assumptions, and to relate to themselves and their lives in analogous terms – in terms, that is, of a problematic of 'the self'. The conceptual dispersion of 'the self' appears to go hand in hand with its 'governmental' intensification.

Have we then, despite the arguments of the critical theorists and philosophers, become 'psychological subjects'? At the end of this series of genealogical studies of the *government* of subjectivity, and of the role of psy within those rationalities and technologies of government, it is time to address the issue of 'subjectivity' more directly. If only, and this will be my argument, to try to pose it differently – not in terms of the effects of 'culture' upon 'the person', or in terms of a 'theory of the subject', but by seeking to characterize the mode of action, as it were, of the diverse psy technologies of subjectification that I have discussed. This will require a detour through some contemporary writings on 'the problem of the subject', before returning, in conclusion, to a consideration of the kinds of creatures we have become.

You are more plural than you think

Perhaps it is Gilles Deleuze and Fèlix Guattari who have articulated the most radical alternative to the conventional image of subjectivity as coherent, enduring, and individualized: "You are a longitude and a latitude, a set of speeds and slownesses between unformed particles, a set of nonsubjectified affects. You have the individuality of a day, a season, a year, a *life* (regardless of its duration) – a climate, a wind, a fog, a swarm, a pack (regardless of its singularity). Or at least you can have it, you can reach it" (Deleuze and Guattari, 1988, p. 162). You can have it – for Deleuze and Guattari, humans, at least along one plane of existence, are more multiple, transient, nonsubjectified than we so often are made to believe. Further, we can act upon ourselves to inhabit such nonsubjectified forms of existence. These nonsubjectified forms they term "haeccities" – modes of individualization that are not those of a substance, a person, a thing, or a subject but of a cloud, a winter, an hour, a date – "relations of movement and rest between molecules or particles, capacities to affect and be affected" (p. 261). Yet opposed to this dimension or 'plane of consistency', which should not be thought of as hidden structure but as an 'immanent' plane consisting only in the distribution and relation amongst its effects, is another plane: that of organization, stratification, territorialization.

> The plane of organization is constantly working away at the plane of consistency, always trying to plug the lines of flight, stop or interrupt the movements of deterritorialization, weigh them down, restratify them, reconstitute forms and subjects in a dimension of depth. Conversely, the plane of consistency is constantly extricating itself from the plane of organization, causing particles to spin off the strata, scrambling forms by dint of speed or slowness, breaking down functions. (p. 270)

If we do not experience and relate to ourselves as movements, flows, decompositions, and recompositions, this is because of the location of humans on this other plane, this plane of organization that concerns the development of forms and the formation of subjects, within assemblages, whose vectors, forces, and interconnections subjectify human being, through assembling us together with parts, forces, movements, affects of other humans, animals, objects, spaces, and places. It is within these assemblages that subject effects are produced, effects of our being-assembled-together. Subjectification is thus the name one can give to the effects of the composition and recomposition of forces, practices, and relations that strive or operate to render human being into diverse subject forms, capable of taking themselves as the subjects of their own and others practices upon them.

No doubt there are many difficulties with these hypotheses that I have taken out of their context and turned to my own account.[1] I am, in any event, less concerned with being 'true to Deleuze and Guattari' – which would be a curious aspiration – than with using their writing as a jumping-off point for my own question: how humans are subjectified, in what assemblages, and how one might conceive of psy as an operative element within them. In an earlier chapter I pointed to the paradoxical position of those who would use a 'theory of the subject' – whose very conditions of possibility lie within a certain historical regime of subjectification – to account for that regime of subjectification itself. Within such arguments, theories of subjectivity are deployed to account for events those theories themselves have helped to produce, through their dissemination through our existence and through their localization within an interiority that they themselves have helped to hollow out. Hence, in the discussion that follows, I would like to propose an approach to the analysis of subjectification that does not utilize a metapsychology to explain how we have become what we are at a particular historical and cultural moment.

'The self' should not be investigated in the terms in which it has historically come to relate to itself, as an enclosed space of human individuality, bounded by the envelope of the skin. "Why should our bodies end at our skin, or include at best other beings encapsulated by the skin? From the seventeenth century until now, machines could be animated – given ghostly souls to make them speak or move or to account for their orderly development and mental capacities. . . . These machine organism relations are obsolete, unnecessary"

(Haraway, 1991, p. 180). Indeed the very idea, the very possibility of a theory of a discrete and enveloped body inhabited and animated by its own soul – *the* subject, *the* self, *the* individual, *the* person – is part of what is to be explained, the very horizon of thought that one can hope to see beyond. If human beings have come to posit themselves as subjects, with a will to be, a predisposition toward being, this does not, as some suggest, arise from some ontological desire but is rather a resultant of a certain history and its inventions (cf. Braidotti, 1994b, p. 160). To write in the spirit of Deleuze would, I think, imply that it was more productive to pose our questions in terms of what humans can do, rather than what they are. Our inquiries would pursue the lines of formation and functioning of an array of historically contingent 'practices of subjectification', in which humans are capacitated through coming to relate to themselves in particular ways: understand themselves, speak themselves, enact themselves, judge themselves in virtue of the ways in which their forces, energies, properties, and ontologies are constituted and shaped by being linked into, utilized, inscribed, incised by various assemblages.

From this perspective, subjectivity – even as a latent capacity of a certain kind of creature – is certainly not to be regarded as a primordial given. Nor is it something that is to be accounted for by 'socialization', by the interaction between a human animal biologically equipped with senses, instincts, needs, and an external, physical, interpersonal, social environment, in which an inner psychological world is produced by the effects of culture upon nature. On the contrary, I suggest that all the effects of psychological interiority, together with a whole range of other capacities and relations, are constituted through the linkage of humans into other objects and practices, multiplicities and forces. It is these various relations and linkages which assemble subjects; they themselves give rise to all the phenomena through which, in our own times, human beings relate to themselves in terms of a psychological interior: as desiring selves, sexed selves, laboring selves, thinking selves, intending selves capable of acting as subjects (see my discussion in Chapter 1, and Rose 1995a, 1995b; cf. Grosz, 1994, p. 116). Subjects, I will argue, might better be seen as 'assemblages' that metamorphose or change their properties as they expand their connections, that 'are' nothing more or less than the changing connections into which they are associated (cf. Deleuze and Guattari, 1988, pp. 8–25). And I shall suggest, as I have throughout this book, that the multitude of lines that have assembled humans into different relations in the twentieth century – the 'rhizomes' that have connected, captured, varied, expanded, diverged, formed points of entry, points of detachment and exit for humans – owe something significant to those concepts, actions, authorities, stratifications, and linkages for which I have used the term 'psy'.

I have tried to demonstrate in previous chapters that psychology, as a body of professional discourses and practices, as an array of techniques and systems of judgment, and as a component of ethics, has a particular significance in relation to contemporary assemblages of subjectification. Psy comprises more than a historically contingent way of representing subjective reality. Psy, in the sense I have given it here, has entered constitutively into critical reflections on the problems of governing persons in accordance with, on the one hand, their nature and truth and, on the other, with the demands of social order, harmony, tranquillity, and well-being. Psy knowledges and authorities have given birth to techniques for shaping and reforming selves assembled together within the apparatuses of armies, prisons, schoolrooms, bedrooms, clinics, and much more. They are bound up with sociopolitical aspirations, dreams, hopes, and fears, over such matters as the quality of the population, the prevention of criminality, the maximization of adjustment, the promotion of self-reliance and enterprise. They have been embodied in a proliferation of social programs, interventions, and administrative projects. In this way, psy has established a variety of 'practical rationalities', has been engaged in the multiplication of novel technologies and their swarming throughout the texture of everyday life: norms and devices according to which the capacities and conduct of humans have been rendered intelligible and judgeable. These practical rationalities are regimes of thought, through which persons can accord significance to aspects of themselves and their experience, and regimes of practice, through which humans can 'ethicalize' and 'agent-ize' themselves in particular ways – as parents, teachers, men, women, lovers, bosses – through their association with various devices, techniques, persons, and objects.[2]

Storying a self

Let us begin with language. In his famous essay on the history of the notion or concept of the self, in which he argued for the recency of the emergence of the category of the self, and the associated cult of the self and respect for the self in law and morality, Marcel Mauss explained that he was not going to discuss language. He believed that there was no tribe or language in which the words 'I' and 'me' did not exist and clearly represent something, and that the omnipresence of the self was also expressed in language by the abundance of positional suffixes that concern the relations in time and space between the speaking subject and that about which they speak (Mauss, 1979b, p. 61). Language itself was here accorded subjectifying effects, even if the subjects so formed did not always reflect upon themselves as subjects in the sense our own culture gives to this term. A different but related argument concerning the subjectifying properties of language was put forward by Emile Benveniste. Benveniste placed great emphasis on the subject-creating properties of personal pronouns, in which the I, as subject of an enunciation, forms a locus of subjectification creating a 'subject position', a place within which a subject can appear (Benveniste, 1971). It was, he argued, through language that humans constituted themselves as subjects, because it was language alone that could establish the capacity of the person to posit him- or herself as a subject, "as the psychic unity that transcends the totality of the actual experiences it assembles and that makes the permanence of consciousness." Subjectivity "is only the emergence in the being of a fundamental property of language" (ibid., p. 224). Language both makes possible, and is made possible by, the fact that each speaker sets themself up *as a* subject by referring to themself as 'I' in their discourse. Pronominal forms are an ensemble of 'empty' signs, without reference to any reality, which become 'full' when the speaker introduces them into an instance of discourse. Yet just because of this, the place of the subject is one that has constantly to be reopened, for there is no subject *behind* the 'I' that is positioned and enabled to identify itself in that discursive space: the subject has to be reconstituted in each discursive moment of enunciation (cf. Coward and Ellis, 1977, p. 133).

For current purposes, however, this emphasis on the subjectifying properties of language conceived of as a grammatical system, a relation among pronouns put into play in instances of discourse, is insufficient. Subjectification can never be purely a linguistic operation. One must agree here with Deleuze and Guattari that subjectification is never a purely grammatical process; it arises out of a "regime of signs rather than a condition internal to language" and this regime of signs is always tied to an assemblage or an organization of power (Deleuze and Guattari, 1988, p. 130). Subjectification, from this perspective, must refer first of all, not to language and its internal properties, but to what Deleuze and Guattari term, drawing upon Foucault, an 'assemblage of enunciation'. In *The Archaeology of Knowledge,* Foucault proposed the term 'enunciative modalities' to conceptualize the forms in which language appears in particular spaces and times, forms that are irreducible to linguistic categories (Foucault, 1972a). Who can speak? From where can they speak? What relations are in play between the person who is speaking and the object of which they speak, and those who are the subjects of their speech? One might think here of the regime that, at any particular time and place, governs the enunciation of a diagnostic statement in medicine, a scientific explanation in biology, an interpretive statement in psychoanalysis, or an expression of passion in erotic relations. These are not put into discourse through the "unifying function of *a* subject," nor do they produce such a subject as a consequence of their effects: it is a matter here of "the various statuses, the various sites, the various positions" that must be occupied in particular regimes if something is to be sayable, hearable, operable: the physician, the scientist, the therapist, the lover (Foucault, 1972a, p. 54). Thus relations among signs are always assembled within other relations: "Only one side of the assemblage has to do with enunciation or formalizes expression; on its other side, inseparable from the first, it formalizes contents, it is a machinic assemblage or an assemblage of bodies" (Deleuze and Guattari, 1988, p. 141).

From this perspective, language itself, even in the form of 'speech', appears as an assemblage of 'discursive practices', from counting, listing, contracting, singing, through the chanting of prayers, to issuing orders, confessing, purchasing a commodity, making a diagnosis, planning a campaign, debating a theory, explaining a process. These practices do not inhabit an amorphous and functionally homogeneous domain of meaning and negotiation among individuals – they are located in particular sites and procedures, the affects and intensities that traverse them are prepersonal, they are structured into variegated relations that grant powers to some and delimit the powers of others, enable some to judge and some to be judged, some to cure and some to be cured, some to speak truth and others to acknowledge its authority and embrace it, aspire to it, or submit to it.

I will return to this argument presently. But in the light of what has been said so far, I want to examine some recent developments within psychology itself that consider subjectification in relation to language and that seek to account for the self in terms of 'narrative': the stories we tell one another and ourselves.

"Not only do we tell our lives as stories, but there is a significant sense in which our relationships with one another are lived out in narrative form" (Gergen and Gergen, 1988, p. 18). For those who argue in this fashion, selves are actually constituted within talk. Language, here, is understood as a complex of narratives of the self that our culture makes available and that individuals use to account for events in their own lives, to accord themselves an identity within a particular story, to attribute significance to their own and others' conduct in terms of aggression, love, rivalry, intention, and so forth. Talk about the self, that is to say, is both constitutive of the forms of self-awareness and self-understanding that human beings acquire and display in their own lives, and constitutive of social practices themselves, to the extent that such practices cannot be carried out without certain self-understandings (Shotter and Gergen, 1989, p. x):

> Instead of assuming that people's relations with nature and with society are unaffected by the language within which they are formulated, we find that these very relations are constituted by the ways of talk informing them, by the forms of accountability by which they are, so to speak, kept in good repair. . . . [I]f we now find ourselves experiencing ourselves as self-contained, self-controlled individuals, owing nothing to others for our nature as such, we need not presume that this is a fixed or 'natural' state of affairs. Rather, it is a form of historically dependent intelligibility requiring for its continued sustenance a set of shared understandings.

Selfhood, and beliefs about the attributes of the self, feelings, intentions, and the like, are understood here as properties, not of mental mechanisms but of conversations, grammars of speaking. They are both possible and intelligible only in societies where these things can properly, grammatically be said by persons about persons. "The task of psychology is to lay bare our systems of norms of representation . . . the rest is physiology" (Harré, 1989, p. 34). Rules of 'grammar' concerning persons, or what Wittgenstein termed 'language games', produce or induce a moral repertoire of relatively enduring features of personhood in inhabitants of particular cultures. "[O]ur understanding and our experience of our reality is constituted for us, very largely, by the ways in which we *must* talk in our attempts . . . to account for it" (Shotter, 1985, p. 168), and we must talk this way because the requirement to meet our obligations as responsible members of a particular society has a morally coercive quality.

These notions of the constitution of the features of personhood within speech are often considered to require a more explicitly 'dialogic' analysis. Such an analysis, it is claimed, can itself serve as a kind of critique of certain ways of talking up the self: the reference to the solitary individual serves misleadingly to locate in the 'I' what is actually a product of a set of relations: "we *talk* in this way about ourselves because we are entrapped within what can be thought of as a 'text', a culturally developed textural resource – the text of 'possessive individualism' – to which we seemingly must (morally) turn, when faced with the task of describing the nature of our experiences of our relations to each other and to ourselves" (Shotter, 1989, p. 136). Historically and culturally developed sense-making procedures, practices, or methods "are made available to us as *resources* within the social orders into which we have been socialized" (ibid., p. 143) and in drawing upon these and using them in their encounters, people come to know themselves as persons of a particular type through an act of mutual recognition. Analysis here thus takes the form of a kind of 'interactional ethnography' of the 'ways of speaking' that are used by people in enacting their social encounters, and in which they mutually construct themselves through the management of meaning.

It is this dialogic character of self-narratives, the fact that they are 'social and not individual', that has recently come to the fore (cf. Hermans and Kempen, 1993). By 'social', as will have already become evident, these authors mean interpersonal and interactional. Thus Mary and Kenneth Gergen argue for the importance of what they term 'self-narratives', culturally provided stories about selves and their passage through their lives that provide the resources drawn upon by individuals in their interactions with one another and with themselves. "Narratives are, in effect, social constructions, undergoing continuous alteration as interaction progresses . . . the self-narrative is a linguistic implement constructed by people in relationships to sustain, enhance or impede various actions. . . . [Self-narratives] are symbolic systems used for such social purposes as justification, criticism and social solidification" (Gergen and Gergen, 1988, pp. 20–1). In explicitly or implicitly organizing their relations to themselves and others in terms of such narratives, a self is, as it were, 'storied forth', the individual choosing among the different narrative forms to which he or she has been exposed.

The 'multiplicity' of the self here is understood to follow from the proposition that "the individual harbors the capacity for a multiplicity of narrative forms" and masters a range of means of making themself intelligible through narratives according to the demands made in the negotiation of social life – for instance, that one makes oneself intelligible as an enduring, integral, or coherent identity (Gergen and Gergen, 1988, p. 35). But "[a]lthough the object of the self-narrative is a single self, it would be a mistake to view such constructions as the product or possession of single selves. . . . in understanding the relationship amongst events in one's life, one relies on discourse that is born of social interchange and inherently implies an audience" (p. 37). This is a sociality that is enhanced by the relational forms and responses that certain ways of talking about the self receive in ongoing interchanges between persons of various types in which individuals negotiate together particular accounts of themselves and others, negotiations that themselves take certain culturally available storied forms.

These studies of 'the self' as constructed in interactional narratives according to available cultural resources certainly pick up on something of considerable significance. If subjectification is analyzed in terms of humans' relations to themselves, as I have argued in Chapter 1, discursively established vocabularies play a significant role in the composition and recomposition of such relations. But analyses conducted under these 'social constructionist' auspices are flawed by the view of language they hold. Language, in such analyses, is seen as 'talk', situationally negotiated meanings between individuals. As 'talk', its analysis follows the banal model of communication, or miscommunication, in which the parties involved, human individuals, utilize various linguistic resources – words, explanations, stories, attributions – to construct messages that will convey intentions, or move, persuade, act upon one another. Such analyses inescapably posit the human agent as the core of sense-making activities, in actively negotiating his or her way through available accounts in order to live a meaningful life. Hence the human being is understood as that agent which constructs itself as a self through giving its life the coherence of a narrative. Evidently, 'the self', simply by virtue of being capable of narrating 'himself or herself' in a variety of ways, is implicitly reinvoked as an inherently unified outside to these communications. One is reminded of Nietzsche's remark that "a thought comes when 'it' wants, not when 'I' want. . . . *It* thinks: but that this 'it' is that famous old 'I' is, to put it mildly, only an assumption, an assertion, above all not an immediate certainty" (Nietzsche, [1886] 1973, pp. 28–9). Yet, for our radical psychologists, it is indeed the familiar old self that is invoked, that comforting 'I' of humanist philosophy, who is the actor interacting with others in a cultural and linguistic context, the person in whom the effects of meaning, communication, take their form, with all the attached assumptions concerning the singularity and cumulativeness of the lived time of consciousness. It is the self of hermeneutics, the self of phenomenology, now being postulated here as the solution to the problem of how it might itself be a possibility.[3]

Of course, it would be absurd to place the analyses produced by linguists like Benveniste in this same hermeneutic camp. Their work refreshes like a glass of iced water when taken after the cloying humanism of 'social constructionists' and demands attention more generous and productive than I have been able to give it here. But nonetheless it is time to question the whole tyranny of 'language', of 'communication', of 'meaning' that has been invoked by the 'social' knowledges for so long in the course of their claims to be distinguished from the 'natural sciences' in virtue of the essential nature of their object. In accounting for our history and our specificity, it is not to the realm of signs, meanings, and communications that we should turn, but to the analytics of techniques, intensities, authorities, and apparatuses. Analyses such as those I have been discussing accord too much to language as communication, and nothing at all to language as assemblage. It might be "relatively easy to stop saying 'I,' but that does not mean that you have gotten away from the regime of subjectification; conversely, you can keep on saying 'I,' just for kicks, and already be in another regime where personal pronouns function only as fictions" (Deleuze and Guattari, 1988, p. 138). If language is organized in regimes of signification through which it is distributed across spaces, times, zones, and strata, and assembled together in practical regimes of things, bodies, and forces, then the 'discursive construction of the self' appears rather differently. Who speaks, according to what criteria of truth, from what places, in what relations, acting in what ways, supported by what habits, routines, authorized in what ways, in what spaces and places, and under what forms of persuasion, sanction, lies, and cruelties? In relation to psy, these are precisely the kinds of questions we must address: the emergence of practices, locales, and enunciative regimes that empower certain authorities to speak of our truth in the language of the psyche; the regimes that constitute authority through a relation with those who are its subjects as patients, analysands, clients, customers; the landscapes, buildings, rooms, arrangements designed for such encounters, from consulting room to hospital ward; the affective vectors of compulsion, seduction, contract, and conversion that connect up the lines.

It is not, that is to say, a question of what a word, a sentence, a story, a book 'means' or what it 'signifies', but rather "what it functions with, in connection with what other multiplicities its own are inserted and metamorphosed" – not its depths or hidden semantic weight, but its 'superficial' connections, associations, activities (Deleuze and Guattari, 1988, p. 4). This is not to turn one's back on language, or on all the instructive studies that have been conducted under the auspices of a certain notion of 'discourse' or have deployed the analytics of rhetoric. But it is to suggest that such analyses are most instructive when they focus not on what language *means* but on what it *does:* what components of thought it connects up, what linkages it disavows, what it enables humans to imagine, to diagram, to hallucinate into existence, to assemble together: sexes with their gestures, ways of walking, of dressing, of dreaming, of desiring; families with their mommies, daddies, babies, their needs and their disappointments; curing machines with their doctors and patients, their organs and their pathologies; psychiatric machines with their reformatory architectures, their grids of diagnosis, their mechanics of intervention, and their notions of cure.[4]

In any event, we should acknowledge that language is by no means primary in the making up of persons. In the first place, language is, of course, more than just 'talk' – thus the significance, which is well recognized, of the invention of writing by means of which humans are capable of becoming 'writing machines' through the training of the hand and eye, the fabrication of instruments such as styli, brushes, pens, a certain set of corporeal habits, a mode of composing and deciphering, a relation to the more or less transportable surface of inscription. In writing, the human being becomes capable of new things: making lists, sending messages, accumulating information from distant locales in a single place and a single plane and comparing, charting changes, differences, and similarities, extending new lines of force (Goody and Watt, 1968; Goody, 1977, pp. 52–111; Ong, 1982). The invention of the printing press makes possible the generalization of 'reading machines' and a multitude of new things become thinkable: novel ways of understanding humans' place in a cosmology through calculations of the movements of the heavenly bodies, for example, or new ways of practicing spiritually in relation to the 'holy book' (Eisenstein, 1979). The invention of techniques of numeracy, or the capacity to calculate, similarly makes humans capable of new things, disciplines thought and self-relations in a distinctive way (foresight and prudence, for example, as one calculates one's future financial situation in the form of a budget), and is similarly dependent upon techniques and devices – machinated assemblages in which the forces of the human are created and stabilized (Cline-Cohen, 1982; cf. Rose, 1991).

Plato, as is well known, expressed severe reservations about writing, conceiving it not only as inferior to the spoken word "written on the soul of the hearer to enable him to learn about the right, the beautiful and the good" but also as destructive of the arts of rhetoric and memory (Plato, *Phaedrus* 278a). But memory should not be counterposed to writing as something immediate, natural, a universal psychological capacity, but seen in terms of what Nietzsche called 'mnemotechnics' (Nietzsche, [1887] 1956, p. 192; cf. Grosz, 1994, p. 131).[5] This term referred to the devices by which one 'burns' the past into oneself and makes it available as a warning, a comfort, a bargaining device, a weapon, or a wound. "Wherever man has thought it necessary to create a memory for himself, his effort has been attended with torture, blood, sacrifice" (Nietzsche, pp. 191–2). Nietzsche's concerns are with the historical varieties of cruel punishment, as examples of the price paid by human beings in order to make them overcome their forgetfulness and "to remember five or six 'I won'ts' which entitled [them] to participate in the benefits of society" (p. 194). It is not a matter, for my purposes, of the validity of Nietzsche's specific genealogical assertions – these are certainly flawed. But the notion of mnemotechnics opens onto a very significant field of investigation for the assembling of subjects. Frances Yates has compellingly shown that memory can be understood as an art or a series of techniques inculcated in the form of particular procedures: an art that was revived and extended in the Middle Ages and involved such techniques as the invention of places or spaces in which items of knowledge or experience were 'placed' and which could be 'retrieved' by the subject taking an imaginary walk among them (Yates, 1966; cf. Hirst and Woolley, 1982, p. 39). The practices of pedagogy have, of course, invented a whole array of other techniques of memory and sought to inculcate them in the schoolrooms that have proliferated throughout the experience of almost all contemporary humans, and which have themselves become infused by psy. But to recognize the technical and practical accomplishment of memory is only a first step: these techniques of memory are not bounded by the envelope of the skin of the subject, let alone by the volume of the brain. Not just the blows, torture, sacrifices, and so forth that Nietzsche discovers as the impure roots of our apparently pure moral nostrums, but also pledges, rituals, songs, writing, books, pictures, libraries, money, contracts, debt, buildings, architectural design, the organization of space and time: all these and much more establish the possibility that a certain more or less imaginary past can be reevoked in the present or the future in particular sites. Memory, that is to say, is itself assembled. One's memory of oneself as a being with a psychological biography, a line of development of emotion, intellect, will, desire, is produced through family photograph albums, the ritual repetition of stories, the actual or 'virtual' dossier of school reports, and the like, the accumulation of artifacts, and the attachment of image, sense, and value to them, and so forth.

Psy, of course, has embraced and deployed memory technologies since at least the time of Mesmer and has been engaged in a whole history of disputation about the status of the memories so produced (Mesmer, [1799] 1957). And memory was central to notions of 'nervous disorder' before Freud announced that hysterics suffered from reminiscences and raised the possibility that the consequences of memory may not distinguish between experience and phantasy. For at least a century, psy claims about memory have been controversial precisely because the memories in question appeared to be the product of its 'unnatural' technologies – of which hypnosis and free association were only two. The contemporary difficulties of psy mnemotechnics are exemplified in what one might term a 'crisis of memory' around the production, through the technologies of psychotherapy, of previously absent memories of childhood abuse – 'false memories', 'recovered memories'.[6] The disputes about this issue reveal, at least in part, the difficulty of recognizing that which is remembered is so only through the engagement of humans within memory technologies. Certain of these technologies, which remain foreign and malign to many cultures, have been 'naturalized' in our own – mirrors, portraits, durable inscriptions (e.g., written script, birthday cards, and letters, which 'stand in' for events past but 'not forgotten'), narrative novels, photographs, now perhaps the video of one's mother's pregnancy and the moment of one's birth. Many of those invented in the genealogy of the psy disciplines – though surprisingly not such memory devices as the medical 'case history' – remain of problematic status, not yet naturalized but still seen as suspect because of their association with the apparently antinatural technology that has forced them into existence. But one is capable of being a 'person-with-memory' only by virtue of 'being-composed-together' with these heterogeneous elements – memory in the sense in which it makes a difference to the ways in which humans act and relate to themselves, is a property of 'remembering machines'.

Memory, numeracy, writing merely exemplify the fact that analyses of language in terms of meaning grant too much autonomy to the semantic and the syntactic, and give too little attention to the situated practices that enjoin, inscribe, incite certain relations to oneself. They ignore the inscription devices, from storybooks to charts, graphs, lists and diagrams, stained glass windows and photographs, the design of rooms, pieces of equipment such as television sets or cooking stoves, cultural technologies that serve as ways of encoding, stabilizing, and enjoining 'being human' that go beyond the envelope of the person, which endure in particular places, practices, rituals, and habits, and which are not localized in particular people or exchanged upon the model of communication.

Thus, while languages, vocabularies and forms of judgment are undoubtedly of immense significance in enjoining and stabilizing certain relations to oneself, they should not be understood as primarily interactional and interpersonal matters. What makes any particular exchange possible arises out of a *regime* of language, embedded within practices that catch up human being in various forms, inscribe, organize, shape, require the production of speech – medical, legal, economic, erotic, domestic, spiritual. But this reference to the practices and assemblages of which language forms a part draws attention to another inescapable weakness of these 'psychological' stories of the narrated self. To the extent that language is situated in such accounts, it is merely in the vague Wittgensteinian shape of 'forms of life' in which 'accountability' functions to make actions possible. Such throwaway references to forms of life are hardly adequate to the task. What needs to be analyzed is the mode of relation to oneself enjoined in definite practices and procedures, links, flows, lines of force that constitute persons, and run across, through, and around them in particular machinations of forces – for laboring, for curing, for reforming, for educating, for exchanging, for desiring, not just for accounting, but for holding accountable. I have examined many of these in previous chapters. As will be evident from those investigations, this is not a call for a more delicate and subtle locating of communication 'in its social context', but for a rejection of the binary form that separates out language only in order to reembed it contextually in a world that is reduced to a kind of cultural background to meaning.

Once technicized, machinated, and located in places and practices, a different image of 'the construction of persons' emerges. Persons, here, function in an inescapably heterogeneous way, as arrangements whose capacities are made up and transformed though the connections and linkages within which they are caught up in particular spaces and places. It is not, therefore, a matter of analyzing the storying forth of the self, but rather, of examining the assembling of subjects: of fighting subjects in the machines of war, laboring subjects in the machines of work, desiring subjects in the machines of passion, responsible subjects in the diverse machines of morality. In each case, the subjectification in question is a product, neither of the psyche nor of language, but of a heterogeneous assemblage of bodies, vocabularies, judgments, techniques, inscriptions, practices.

Imaginary anatomies

I suggested earlier that we might produce more in terms of intelligibility if we considered the question of subjectification less in terms of what kind of subject is produced – a self, an individual, an agent – and more in terms of what humans are enabled to do through the forms into which they are machinated or composed. What humans are able to do is not intrinsic to the flesh, the body, the psyche, the mind, or the soul; it is constantly shifting and changing from place to place, time to time, with the linking of humans into apparatuses of thought and action – from the simplest connection between one organ or body part and another in terms of an 'imaginary anatomy' to the flows of force made possible by the links of an organ with a tool, with a machine, with parts of another human being or beings, within an assembled space such as a bedroom or a schoolroom. From this perspective, the questions to be addressed concern not 'the constitution of the self' but the linkages established between the human and other humans, objects, forces, procedures, the connections and flows made possible, the becomings and capacities engendered, the possibilities thus foreclosed, the machinic connections formed that produce and channel the relations humans establish with themselves, the assemblages of which they form elements, relays, resources, or forces (cf. Grosz, 1994, p. 165; Deleuze and Guattari, 1988, p. 74).

In thinking this way, we might read in reverse, as it were, the many recent texts that seek to ground their analytics of power relations and forms of knowledge upon 'the body'. Human corporeality, it is often suggested, can provide the basis for an account of subjectification, the constitution of desires, sexualities, and sexual differences, the phenomena of resistance and agency. Human beings are, after all, embodied, such arguments assert, despite all the attempts of philosophers since 'the enlightenment' to portray them as creatures of reason, and to assert that this capacity to reason removes humans – or at least male humans – almost entirely from their characteristics as creatures. And although accepting that corporeality does not give any essential or stable form to subjectivity, how could one deny the claim in such analyses that it is upon the raw material of 'the body' that culture works out its constitution of subjectivity? While abjuring all forms of psychological essentialism, how could one disagree with the assertion that the forms of subjectivity are irretrievably marked by the biological facticity of sexed bodies, of infant bodies that are incapable of self-maintenance, of all bodies that eat, drink, copulate, defecate, deteriorate, and die (e.g., Butler, 1990, 1993). This ambivalence is encapsulated in Braidotti's assertion that "the starting point for feminist redefinitions of subjectivity is a new form of materialism that places emphasis on the embodied and *therefore* sexually differentiated structure of the speaking subject" (1994a, p. 199, emphasis added). And such is the apparent compulsion of such a way of thinking that even such an antinaturalist writer as Elizabeth Grosz, who wishes to question all essentialisms and binaries, suggests that 'the body' is the material upon which culture, history, and technique writes, and hence "the bifurcation of sexed bodies . . . is . . . an irreducible cultural universal" (Grosz, 1994, p. 160).

But 'the body' is itself a historical phenomenon. Our current image of the lineaments and topography of 'the body' – its organs, processes, vital fluids, and flows – is an outcome of a particular cultural, scientific, and technical history. The properties of 'the body' – walking, smiling, digging, swimming – are not natural but technical achievements (Mauss, 1979a). Even the apparent naturalness of the limits and boundaries of 'the body', which appear to define as if inevitably the coherence of an organic unity, is both recent and culturally specific (Foucault, 1973; cf. Grosz, 1994, on the history of the notion of 'body image'). And as for the 'two sexes', so many historical studies show how diverse is this apparently imutable division, what intellectual labors have been entailed in stabilizing it in a doubled nature of the male and female body, in making our sexual desire our hidden secret, connecting up pleasure, sex, will, knowledge, reproduction, and companionship in the 'cyborg sexuality' that we have come to inhabit it as our truth (e.g. Feher, 1989a, 1989b, 1989c; Laqueur, 1990; Brown, 1989; cf. Valverde, 1985, on our fabrication as sexually desiring subjects). Hence much recent emphasis in feminist writing on the body and embodiment preserves the very analytic it seeks to subvert, displacing the 'enlightenment' normalization of the properties of reason and abstraction by merely reversing the old trope that females are more bodily, more fleshly, but nonetheless retaining the flesh as the governing perspective of feminist reason. But bodies are always 'thought bodies' or 'bodies-thought' – and perhaps some day we will look back on the 'sex-thought-body' that has so exercised our own century, our own repetitive and wearying anxiety about our sexual bodies, our commitments to the difference of gender that marks us so indelibly, the transgressive forces and restorative powers of the sexual and all the rest, with a certain wry amusement (cf. Foucault, 1979b, pp. 157–9).

Let us therefore abandon this 'fleshism' of the body once and for all.[7] 'The body' is far less unified, far less 'material' than we often think. Perhaps, then, there is no such thing as '*the body*': a bounded envelope that can be revealed to contain within it a depth, and a set of lawlike operations. What we are concerned with, at least in the kinds of investigation being undertaken here, are not bodies, but the linkages established between particular surfaces, forces, and energies. Rather than speak of 'the body', we need to analyze just how a particular body-regime has been produced, the channeling of processes, organs, flows, connections, the alignment of one aspect with another. Instead of 'the body', then, one has a series of possible 'machines', assemblages of various dimensions of humans with other elements and materials: connected to books to form a literary machine, to tools to form a work machine, to goods to form a consuming machine, and so forth. The body is thus "not an organic totality which is capable of the wholesale expression of subjectivity, a welling up of the subject's emotions, attitudes, beliefs, or experiences, but is itself an assemblage of organs, processes, pleasures, passions, activities, behaviors, linked by fine lines and unpredictable networks to other elements, segments and assemblages" (Grosz, 1994, p. 120). And organs themselves are 'tactile', the eye, the nose, the ear, the touch conjoining thought and object in sensuous relations of contact, exchange, and interpenetration, creating a multitude of new sensoria through each "flashing moment of mimetic connection, no less embodied than it is mindful, no less individual than it is social" (Taussig, 1993, p. 23: while the argument is Taussig's, he is discussing here the work of Walter Benjamin).

Our regime of corporeality thus should itself be regarded as the unstable resultant of the assemblages within which humans are caught up, which induce a certain relation to ourselves *as embodied,* which render the body organically unified, traversed by vital processes, which differentiate – today by sex, for much of our history by 'race' – which accord it a depth and a limit, equip it with a sexuality, establish the things it can and cannot do, define its vulnerability in relation to certain dangers, make it practicable in order to bind it into practices and activities (on 'the woman's body' see, e.g., Laqueur, 1990, Duden, 1991; on the racialized body, see Gilman, 1985). Deleuze's question, which for him was Spinoza's question, "What is a body capable of?" – what can it do, what affects can it have, how do these strengthen, weaken, capacitate it in different ways, multiply it, and metamorphose it? – is a starting point (Deleuze, 1992b, chap. 14). But this only as long as we are clear that a body is not 'the body', but merely a particular relationship capable of being affected in particular ways. It is a matter of organs, muscles, nerves, tracts that are themselves swarmings of cells in constant interchange with one another, linking and detaching, dying, reconfiguring, connecting and combining, where the outside of one is simultaneously the inside of another. And it is also a matter of brains, hormones, chemical molecules that connect and transform the capacities of various parts – exciting them, coordinating them, fusing them, or disengaging them.

These assemblages are not delineated by the envelope of the skin, but link up 'outside' and 'inside' – visions, sounds, aromas, touches, collections together with other elements, machinating desires, affections, sadness, terror, even death. Consider what diverse machinations of the body are capable of: the bravery of the warrior in battle, the tenderness or violence of the lover, the endurance of the political prisoner under torture, the transformations effected by the practices of yoga, the experience of voodoo death, the capacities of trances to render organs capable of withstanding burning or recovering from wounds. These are not properties of 'the body' but machinations of the 'thought body' whose elements, organs, forces, energies, passions, dreads are assembled through connections with words, dreams, techniques, chants, habits, judgments, weapons, tools, groups. This is not to suggest that humans could be angels, fly out of windows, or wriggle down earthworm burrows. But it is to suggest that 'materialist' appeals to corporeality as the 'material' upon which culture works are not very good to think with. Bodies are capable of much, at least, in part, in virtue of their 'being thought' and we do not know the limits of what is possible for such thought-body-machines to do.[8] If we have become psychological creatures, this not because of a givenness of an interior, nor because of the meanings of a culture, but because of the ways in which, in so many locales and practices, psy vectors have come to traverse and link up these machinations.

Two metaphors for the machinations of subject-bodies have recently been proposed: performativity and inscription. Judith Butler proposed the notion of performativity in developing an analysis of the construction of 'gender identity', which does not posit any essential or pregiven subject lying behind its actions. For Butler, one needs "no theory of gender identity behind the expressions of gender . . . identity is performatively constituted by the very 'expressions' that are said to be its results" (Butler, 1990). Her notion of performativity here draws upon Austin and Derrida to argue that gender is the outcome of performative acts. "A performative act is one which brings into being or enacts that which it names, and so marks the constitutive or productive power of discourse. . . . For a performative to work, it must draw upon and recite a set of linguistic conventions that have traditionally worked to bind or engage certain kinds of effects" (Butler, 1995, p. 134). Gender, thus, is a phantasy "instituted and inscribed on the surface of our bodies," constituted through the effects of signification engendered through the performances of language (1990, p. 136). But this notion of performativity limits itself through maintaining its emphasis on the linguistic. Take this argument about the performance of woman-ness, which I owe to Susan Bordo (Bordo, 1993, p. 19):[9]

> Sit down in a straight chair. Cross your legs at the ankles and keep your knees pressed together. Try to do this while you're having a conversation with someone, but pay attention at all times to keeping your knees

> pressed tightly together. . . . Run a short distance, keeping your knees together. You'll find you have to take short, high steps. . . . Walk down a city street. . . . Look straight ahead. Every time a man walks past you, avert your eyes and make your face expressionless.

'Gendering', as Butler recognizes along with many others, is a matter of a meticulous and continually repeated prescription of the deportment, appearance, speech, thought, passion, will, intellect in which persons are assembled by being connected up not only with vocabularies but also with regimes of comportment (walking, looking, gesturing), with artifacts (clothes, shoes, makeup, automobiles, cooking pots, writing implements, books), with spaces and places (classrooms, libraries, railway stations, museums), and the objects that inhabit them (desks, chairs, books, platforms, display cabinets). Performativity, at least in the sense in which it is construed in terms of the model of linguistic utterance, citations, conventions, is a rather misleading image for thinking such an assembling of the person: it is necessary to insist that we are not 'constituted by language'.

Nor is a different linguistic image, that of writing or inscription, sufficient. This notion is used both by Butler and by Grosz to render the relation between, on the one hand, the body and its surfaces as that which is marked, inscribed, engraved and, on the other, "the tracing of pedagogical, juridical, medical and economic texts, laws and practices onto the flesh to carve out a social subject as such, a subject capable of labor, of production and manipulation, a subject capable of acting as a subject and, at the same time, capable of being deciphered, interpreted, understood" (Grosz, 1994, p. 117). Rather than an analytic of inscription, in which culture is written on the flesh, I think it is more useful to think in terms of technology. Indeed, as I have suggested, language, writing, memory can themselves be seen to be elements in a technics, each entailing truths, techniques, gestures, habits, devices assembled through training and embedded in more or less enduring associations. Practices of subjectification might be best understood in terms of the complex interconnections, techniques, and lines of force between heterogeneous components that incite, make possible, and stabilize particular relations to oneself in specific sites and locales. Technologies of subjectification, then, are the machinations, the being-assembled-together with particular intellectual and practical instruments, components, entities, and devices that produce certain ways of being-human, territorialize, stratify, fix, organize, and render durable particular relations that humans may truthfully establish with themselves.

There is no need to posit any 'propulsive medium' behind all these technologies, no primordial force or desire that courses through these assemblages and makes it possible for them to move, act, change, resist, mutate. The 'question of agency' as it has come to be termed, poses a false problem. To account for the capacity to act one needs no theory of the subject prior to and resistant to that which would capture it – such capacities for action emerge out of the specific regimes and technologies that machinate humans in diverse ways (here I find myself in agreement with Butler, 1995, p. 136). The heterogeneity of these practices and techniques – their multiple interconnections, alliances, conflicts, and divergences, the different promises they hold out, and the variable demands they make of human being – can produce all the effects of resistance, appropriation, utilization, transformation, and transgression that theorists of the postmodern have highlighted, without the need to invoke a unifying conception of 'human agency'. To put it another way, agency is itself an effect, a distributed outcome of particular technologies of subjectification that invoke human beings as subjects of a certain type of freedom and supply the norms and techniques by which that freedom is to be recognized, assembled, and played out in specific domains. Indeed, as I have argued in previous chapters, over the past century, psy has had a very particular role in creating the conditions for the emergence of the capacity to relate to oneself as certain types of agent – as 'characters', for example, in the nineteenth century, with nervous functions, which, as shaped by the effect of habit and influence upon one's constitution, produced impulsiveness or restraint depending on whether one was a man or a woman, a master or a mistress, a casual laborer, clerk, or manservant (cf. Smith, 1992, chap. 1); as 'personalities' over the course of the twentieth century, a type possessed of certain traits, manifested in the ways in which one reacted to experience, expressed one's feelings, assembled oneself together with artifacts, tastes, forms of dress, styles of gesture and expression; as 'free agents' of choice and self-actualization, in the second half of the twentieth century, champing against all the machines that would machinate us as good subjects of bureaucracy and conformity, that would lower our self-esteem and hinder our self-realization.

For our own culture, of course, agency is part of an 'experience' of internality – it appears to well up and rise out of our depths, our inner instincts, desires, or aspirations. No doubt it has not always been so. E. A. Dodds's classic account of Homer's *Iliad* and *Odyssey* suggests that the Homeric portrayal of humans is more than a matter of aesthetic convention: humans, for Homer, were dispersed assemblages of *psyche* (soul), *thumos* (will), and *noos* (intellect), each with its own independent mode of operation, and action was not understood in terms of any internal faculty of agency, but in terms of such forces as *ate* which compelled a person to a particular course of action through the intervention of gods, fates, furies, dreams, and visions (Dodds, 1973; cf. Hirst and Woolley, 1982). Such examples could, of course, be multiplied: of the explanatory powers of the voices of deities or devils, the motivating effects of shamans and rituals, closer to home, perhaps, the consequences of the crowd or the mob in sweeping up the individual into a new many-headed agent with a single, if malign, will (on the crowd, see Chapter 6). Agency is, no doubt, a 'force', but it is a force that arises not from any essential properties of 'the subject' but out of the ways in which humans have been-assembled-together.

Folding souls

If, today, we live out our lives as psychological subjects who are the origins of our actions, feel obliged to posit ourselves as subjects with a certain desiring ontology, a will to be, it is on account of the ways in which particular relations of the exterior have been invaginated, folded, to form an inside to which it appears an outside must always make reference. Once more, it is Deleuze who has reflected most instructively on a philosophy of the fold (Deleuze, 1992a, 1992b, and see especially the use of this notion in his discussion of subjectification in his book on Foucault: Deleuze, 1988, pp. 94–123). "[W]hat always matters is folding, unfolding, refolding" (Deleuze, 1992a, p. 137). The concept of the fold can give rise to a generalizable diagram for thinking of relations, connections, multiplicities, and surfaces – their formation of depths, singularities, stabilizations. This diagram of the fold describes a figure in which the inside, the subjective, is itself no more than a moment, or a series of moments, through which a 'depth' has been constituted within human being. The depth and its singularity, then, is no more than that which has been drawn in to create a space or series of cavities, pleats, and fields, which only exist in relation to those very forces, lines, techniques, and inventions that sustain them.

The languages, techniques, institutional sites, and enunciative relations of clinical medicine introduced deep foldings into the body, the inside of this outside, the inside as an operation of the outside as Deleuze suggests in his discussion of Foucault's archaeology of the clinical gaze. Or again, in relation to the ethical techniques introduced by the Greeks, these must be understood "in the sense in which *the relation to oneself* assumes an independent status. It is as if the relations of an outside folded back upon themselves to create a doubling, to allow a relation to oneself to emerge, and to constitute an inside which is hollowed out and develops its own unique dimension" (Deleuze, 1988, p. 100). Once this new dimension is established, the subject is assembled in new ways, in terms of a problem of 'self-mastery': bringing to bear upon oneself – this inside acting upon itself – the power that one brings to bear upon others. In the same process the power that is brought to bear upon others is refigured as a power relation between this inside of oneself and this inside of the other.

This singularized and folded inside is thus inevitably stabilized, not in relation to a domain of psychological processes, but in relation to a configuration of forces, bodies, buildings, techniques that hold it in place. For the Greeks this comprised the whole apparatus of ethical formation established within the city, family relations, tribunals, games of power and of leisure, erotic relations and so forth within which those males who would exercise power were assembled. "This is what the Greeks did: they folded force, even though it still remained force. They made it relate back to itself. Far from ignoring interiority, individuality or subjectivity they invented the subject, but only as a derivative or the product of a 'subjectification'" (Deleuze, 1988, p. 101). This relation to oneself, this folding that produces the effects of subjectification, is not passive. Again as Deleuze points out, it is created only through *practicing it,* through carrying it out, through engaging with the techniques of government of the body, control of diet, techniques of sexuality, styles of games and sports, oration, display in public, and so forth. And the Greeks, while inventing a particular formulation of this dimension of 'being's relations to itself', were by no means the last – nor probably the first – to do so; rather, what they exemplify is one particular form of a more general relation, a relation within which subjectification is always a matter of folding. The human is neither an actor essentially possessed of agency, nor a passive product or puppet of cultural forces; agency is produced in the course of practices under a whole variety of more or less onerous, explicit, punitive or seductive, disciplinary or passional constraints and relations of force. Our own 'agency' then is the resultant of the ontology we have folded into ourselves in the course of our history and our practices. For all the desires, intelligences, motivations, passions, creativities, will-to-self-realization, and the like folded into us by our psychotechnologies, our own agency is no less artificial, no less fabricated, no less unnatural – and hence no less real, effective, confused, technical, machine-dependent – than the problematic agency of the robots, replicants, and monstrous symbioses that Donna Haraway uses to think of our existence: cyborgs, hybrids, mosaics, chimeras (Haraway, 1991, pp. 171–2).

But what is it that is folded? No doubt it is true that for Deleuze what is folded is always some 'force'. Perhaps for our own purposes we might address this question somewhat modestly. In the previous chapters I have used the term 'authority' for the foldings that make a difference. Of course, this merely names a field but does not define or delimit it in principle; the point is that *anything* might have authority. But at any one time and place, not everything does. An analysis here should be of the rarity of authorities in actuality and not of their infinite components and possibilities. It is not into anything that persons may be assembled at any one time and place, and the vectors that are folded, have limits that are not ontological but historical. What is infolded is composed of anything that can acquire the status of authority within a particular assemblage. Machinations of learning, of reading, of wanting, of confessing, of fighting, of walking, of dressing, of consuming, of curing enfold a certain voice (that of one's priest, one's doctor or one's father), a certain incantation of hope or fear (you can become what you want to be), a certain way of linking an object with a value, sense, and affect (the 'Italian-ness' that Barthes so wonderfully reveals within Panzani pasta or perhaps the 'self-control' manifested by the sculpted body of the 'postmodern woman'), a certain little habit and technique of thought (bite the bullet, look before you leap, self-control is everything, it is good to share one's feelings), a certain connection with an authoritative artifact (a diary, a dossier, or a therapist).

Foucault, as we saw earlier, suggested that ethical technologies might be analyzed along four axes (see my discussion in Chapter 1); Deleuze transcribes each of these four axes by means of the concept of folding (Deleuze, 1988).[10] The first, he suggests, concerns the *aspects* of human being that are to be surrounded and enfolded – the body and its pleasures for the Greeks, the flesh and desires for the Christians, the self and its aspirations, perhaps, for our own period. The second, the relation between forces, concerns the *rule* according to which the relation between forces becomes a relation to oneself – a rule that might be natural, divine, rational, aesthetic, and so on. It thus is always associated with a particular authority – that of the priest, the intellectual, the artist; in our own day, perhaps the rule veers between the psychotherapeutic and the stylistic, each associated with different authorities. The third, the fold of knowledge or fold of *truth,* arises in that each relation to oneself is organized on this axis of the subjectification of knowledge, and hence the relation of our being to truth, whether that truth be theological, philosophical, or psychological. The fourth fold (here Deleuze refers to Blanchot's notion of an interiority of expectation) is the fold of hope – for immortality, eternity, salvation, freedom, death, or detachment. And subjectification, then, is the interplay of the multiple variability of these folds, of their diverse rhythms and patterns. "And what can we ultimately say about our own contemporary folds and our modern relation to oneself? *What are our four folds?*" (Deleuze, 1988, p. 105). The essays in this book have, in their different ways, been attempts to answer that question. It is nonetheless appropriate to conclude with some further reflections on the part that the psychosciences and techniques play within these foldings.

Psychologics of subjectification

I have suggested that psy, as I have defined it in these essays, plays a constitutive role in our 'four folds', in complex and variable relations with other vectors, of course, but even then overlaying, infusing, and investing them, such that even aesthetic, spiritual, economic, financial, 'life-style', or erotic ethics are suffused with psy in their enunciative regimes, technologies, modes of judgments, and enactments of authority. Without attempting to repeat or summarize my substantive arguments in the various studies in this book, let me trace out some of the features of these psy foldings.

The aspect of human being that is surrounded and enfolded in so many contemporary assemblages of subjectification is neither body/pleasure nor flesh/desire but self/realization. A psy ontology has come to inhabit us, an inescapable interiority that hollows out, in the depths of the human, a psychic universe with a topography that has its own characteristics – its planes and plateaus, its flows and precipitations, its climates and storms, its earthquakes, volcanic eruptions, warmings and coolings. Of course, the charting of this psy universe is incomplete and disputed, the maps reminiscent of those that guided seafarers in an earlier age; where some report sightings of instincts, inherited characteristics, and predispositions, others have found repressions, projections, and phantasies, others have seen the internalization of social expectations, and others observed merely the inscription of a regime of behavioral rewards and punishments. The dynamics of this ontology are contested, whether they be the processes of self-esteem and self-abnegation, of stress and fulfillment, of desire and frustration, of anxieties and phobias, of the sadistic involutions of internal objects. But these dynamics are assembled through vectors that run across and through the envelope of the skin. Indeed 'the body' is itself now seen as less a corporeal given than as an organic complex whose properties are marked by this psy interior – body image, psychosomatics, the cancer-prone personality, fatness or thinness as manifesting the cravings for love of an inner self, fitness as a kind of psychic economy of self-esteem and empowerment. Inculcation, emulation, mimesis, performance, habituation, and other rituals of self-formation hollow out and shape this 'internal' space in a psy form.

Human ontology is thus established in part through constitutive connections with the psy technologies that imagine it and act upon it. These activate something that Michel Taussig has revealingly examined in terms of 'mimesis' – the becoming enacted in the continuous interplay between the copy and the copied (Taussig, 1993). The copy, here, comprises both a 'representation' – picture, artifact, object, gesture, dance, model, diagram – and a form of being. "Between photographic fidelity and fantasy, between iconicity and arbitrariness, wholeness and fragmentation, we thus begin to sense how weird and complex the notion of the copy becomes" (Taussig, 1993, p. 17). The multitude of little flashes that Taussig terms mimesis enfold certain 'ways of being' into us – not just 'descriptions or stories', not just through 'rewards and punishments' (as if it was ever very clear which was what), but through mime and mimicry, through emulation and bricolage, through both copying and differing. For our purposes, then, the mimetic dimension of psy can be seen in such devices as manuals of advice on self-improvement, self-esteem, and self-advancement; the psy patterns forced into visibility in all the consulting rooms and counseling sessions; the models and simulacra of desirable selves that serve as mirrors to reactivate and reflect back fabrications of subjectivity to which one might aspire; the pictures of normal selfhood – the normal child, the normal mother, the normal girl, the normal adolescent, the normal patient, worker, or manager – deployed in every practice imaginable; the connections established with oneself through cultural technologies of photography, film, and advertising: a multiplicity of mimetic machines. The injunction to be a certain kind of self is always conducted through operations that distinguish at the same time as they identify (see Taussig, 1993, again on this). To be the self *one is* one must not be the self *one is not* – not *that* despised, rejected, or abjected soul. Thus becoming oneself is a recurrent copying that both emulates and differs from other selves. Today the pertinent features of mimesis and alterity are established in the vectors of life-styles, sexualities, personalities, aspirations.

To speak of the folding of this psy ontology into humans is to gesture – at this stage it can be no more than that – to the processes that have hollowed out an interior through the enfolding of psy components that have been distributed across these devices and technologies. This psy space is composed of a complex admixture of elements from psychological research into humans and animals, stories and fabulations, autobiographies and case histories. It is 'fictional' only in the sense that psy 'invents' and reinvents imagined worlds in the quest for what it takes as its premise: that a real world inhabits our being as humans. (cf. Haraway, 1989). And though it is true, no doubt, that the characteristics of this folded world are as crumpled, twisted, ragged, and frayed as the materials of which it is made, nonetheless our relations to ourselves have been, for at least a century, irrevocably marked by our fold of the self, for it is this name that our age has given to the troubled universe within which all the differences of humans will be registered, located, explained, and acted upon.

I have argued in these studies that at least one key dimension of the fold of authority today can be termed 'therapeutic': it is according to a therapeutic rule that lines of force are bent around into a self-shaped space in our existence and experience. Therapeutic, here, not in the sense of a privilege accorded to 'psychotherapy' itself, or even simply in the terms of the proliferation of the branches and varieties of psy – forensic psychologists with their offender and victim profiling, sports psychologists with their mental exercises for success on field or track, organizational consultants with their protocols of increasing productivity and harmony through acting upon the self-actualizing propensities of employees and the like. Therapeutic, rather, in the sense that the relation to oneself is itself folded in therapeutic terms – problematizing oneself according to the values of normality and pathology, diagnosing one's pleasures and misfortunes in psy terms, seeking to rectify or improve one's quotidian existence through intervening upon an 'inner world' we have enfolded as both so fundamental to our existence as humans and yet so close to the surface of our experience of the everyday. It is this therapeutic relation to ourselves, and the authoritative components of this relation that have multiplied in our present, a multiplication of the relays between the authorities that speak the truths of ourselves and the ways in which we act upon our own existence, in the understanding, planning, and evaluation of our everyday passions, fears, and hopes. The self is produced in the practicing of it, hence produced as an interiority that is complex, contested, and fractures, through the intersection of the multitude of activities and judgments that one brings to bear upon oneself in the course of relating to one's existence under different descriptions and in relation to different images or models, the sanctions, seductions, and promises under which one accords these therapeutic ways of practicing subjectivity a value and an authority.

And what of the fourth fold – what can we hope for? What we fold, what folds us, is an aspiration as pathetic as it is touching: no more, it seems, than to maximize our life-styles and fulfill ourselves as persons through our relations with other persons – our lovers, our children, our mothers and fathers, our communities. To this hope we have given the name of 'freedom'. This hope is not one of liberation for the world and its mundane cares, sorrows, and obligations – 'turn on, tune in, and drop out'. Nor is it a freedom from the bonds of servitude and subjection: 'free at last, free at last, thank God almighty, free at last'. Rather, the chimes of a rather different freedom flash in our dreams: a mode of being in the world in which we would accord value to our lives to the extent that we are able to construe them as the expression of a personal autonomy. And this autonomy is conceived in terms that are simultaneously political (free to choose) and psychological (free to choose in the name of ourselves and not in the name of our subordination to the authority of another, in relation to the shadow cast by our internalized parents, or the restrictions imposed by our fear of freedom itself). A laudable aspiration? No doubt, but it is one that does not exist in a relation of externality with our anxieties and our disappointments: this dream of freedom constitutes the very ways in which we codify and experience ourselves, and the ways in which we divide ourselves from that in ourselves, and from that in others, that does not accord with this dream or which fails by its principles.

The psy effect

To research these hypotheses more directly, we might begin by establishing some kind of topography of psy spaces, of the practices or assemblages in which our subjectivity is machinated. One might term this the 'where' of psy: its territorialization. In the course of the studies in this volume we have identified a whole variety of assemblages within which such a territorialization has been organized: desiring machines, laboring machines, pedagogic machines, punitive machines, curative machines, consuming machines, war machines, sporting machines, governing machines, spiritual machines, bureaucratic machines, market machines, financial machines. This is not to assert the dominance of psy in our experience, for could not the same be said of, for example, the languages, images, techniques, and seductions of economics? Nor is it to identify an external 'cause' of all these transformations and mutations that have come to permeate so widely across our existence. But it is to register this 'psy effect' in the sense in which Deleuze understands the notion of an effect, such as the Kelvin effect or the Compton effect, as deployed in scientific discourse: "An effect of this kind is by no means an appearance or an illusion. It is a product which spreads or distends itself over a surface; it is strictly co-present to, and co-extensive with, its own cause, and determines this cause as an immanent cause, inseparable from its effects" (Deleuze, 1990b, p. 70, quoted in Burchell et al., 1991, p. ix). The 'psy effect', that is to say, is not to be identified with a particular cause, but rather delineated by describing the ways in which human existence is rendered intelligible and practicable under a certain description in a whole multiplicity of little 'ethical scenarios' that permeate our experience.

By 'ethical scenarios' I mean the diverse apparatuses and contexts in which a particular relation to the self is administered, enjoined, and assembled, and where therapeutic attention can be paid to those who are rendered uneasy by the distance between their experience of their lives and the images of freedom and selfhood to which they aspire. This is partly a matter of the shaping of space itself. We have many instructive studies of 'disciplinary' architecture, the relations of bodies, gazes, and activities in the machines of morality invented in the nineteenth century: prisons, schools, asylums, reformatories, washhouses, and the like (Markus, 1993; cf. Rose, 1995a). But, with the exception of the attention that authors have recently directed toward shopping, the department store, and the mall, we have fewer studies of the 'seductive architecture' of our own times (on spaces of consumption, see Bowlby, 1985, and Shields, 1992; see also the interesting discussion in Eräsaari, 1991). This would require us to go beyond the tutelary spaces in the school, the courts, the visit of the social worker, the doctor's surgery, the ward group of the psychiatric hospital, the interview with the personnel officer. It would require us to examine also the infusion of psy into the configuration of the home, the gym, the analyst's consulting room, the therapeutic group, the counseling session, the marriage guidance encounter, the radio phone-in. Further, a topography of ethical scenarios would need to examine the spatial and material arrangements put in place by the cornucopia of courses and training experiences that seek to instrumentalize a new psychological conception of human relations. Of particular significance here would be the way in which collections of persons in space and time have been reconstrued as groups traversed by unconscious forces of projection and identification, allowing not only a new dimension for the explanation of collective troubles but a new range of techniques – from T-groups to group therapy – for managing them therapeutically. A multitude of scenarios has been invented for therapeutic engagement with the human subject, an array of locales for cure, reform, advice, and guidance has been transformed according to a 'psy effect'.

What is acted upon? What lines, forces, surfaces, or flows of human being are caught up in these machines? Desires? Yes: undoubtedly one vector of our contemporary relation to ourselves passes through the flows of drives, phantasies, repressions, projections, identifications, and the impulses to speech and conduct that are established within this desiring ontology. But, as I have suggested, we would be wise to avoid constructing some metaphysics of desire, or at least to leave this project to our philosophers. For the genealogist, desire is only one of the vectors of the contemporary psychological machination of human being, of our present 'psy effect'. One would want to stress also the vectors that flow around the superficiality of 'behavior' itself – the pedagogies of social skills and life-style and all the behavioral technologies to which they have given rise. Perhaps equally significant within the new ethical obligations of personal fulfillment has been the new relation of self-to-self exemplified by the notion of self-esteem: "an innovation which transforms the relation of self-to-self into a relationship that is governable" (Cruikshank, 1993), in the course of which a whole panoply of psy techniques have been deployed – coaching in a new vocabulary of self-respect, exercises involving the narrativization of one's life in various therapeutic, pedagogic, or intimate scenarios. Further, despite not so directly appearing to implicate a psy ontology, one would need to examine the techniques of composing and adorning the flesh – styles of walking, dressing, gesture, expression, the face and the gaze, body hair and adornment – a whole machination of being in terms of the look that operates in terms of a relation between the outer and the visible and the inner and invisible. For this relation too, over the course of the twentieth century, has been composed and characterized through cultural technologies of advertising and marketing that have deployed psy devices for understanding and acting upon the relations between persons and products in terms of images of the self, its inner world, and its life-style. Overarching all their differences, contemporary techniques of subjectification operate through assembling together, in a whole variety of locales, an interminably hermeneutic and subjective relation to oneself: a constant and intense self-scrutiny, an evaluation of personal experiences, emotions, and feelings in relation to psychological images of fulfillment and autonomy.

In all these diverse machinations of being, in all these heterogeneous assemblages, a number of themes recur: choice, fulfillment, self-discovery, self-realization. Contemporary practices of subjectification, that is to say, put into play a being that must be attached to a project of identity, and to a secular project of 'life-style', in which life and its contingencies become meaningful to the extent that they can be construed as the product of personal choice. It would be foolish to claim that psychology and its experts are the origin of all these subjectifying machines – it is rather a matter of how assemblages of passion and pleasure, of labor and consumption, of war and sport, of aesthetics and theology, have accorded to their subjects a psychological form. The essays in this book have simply made a start at tracing out the ways in which psychological modes of explanation, claims to truth, and systems of authority have participated in the elaboration of moral codes that stress an ideal of responsible autonomy, in shaping these codes in a certain 'therapeutic' direction, and in allying them with programs for regulating individuals consonant with the political rationalities of advanced liberal democracies.

Unbecoming selves

In the course of these essays I have suggested that one of the intriguing and possibly hopeful features of our current ethical topography is the heterogeneity of the territory mapped out by machinations of the self, the variety of attributes of the person that they identify as of ethical significance, and the diverse ways of calibrating and evaluating them that they propose. Nonetheless, it is important to recognize simultaneously that this ethical territory is not a free space: persons' relations to themselves are stabilized in assemblages that vary from sector to sector, operating via different technologies depending upon one's identification as adjusted or maladjusted, normal or pathological, lawbreaker or honest citizen, man or woman, rich or poor, black or white, employed or unemployed, operating under different forms of authority in the prison and the factory, in the supermarket and the hair stylist, in the bedrooms of the conjugal home and the brothels of the red-light districts, in the new territories of exclusion and marginalization brought into existence by the fragmentation of the social. But this is not to say that the psy effect I have been tracing is confined to a cultural elite. New modes of subjectification produce new modes of exclusion and new practices for reforming the persons so excluded: as, for example, in the deployment of behavioral technologies so widely used in practices of reform that seek to 'empower' their subjects and restore them to the status of freely choosing citizens (Baistow, 1995). New psy models of personhood, and the ethical regimes to which they are attached have no intrinsic political character: they have a versatility, which enables them to multiply, to proliferate, to be translated and utilized in ways that are not given by an internal logic, be it one of emancipation or domination.

Nonetheless, while I have stressed the heterogeneity of the foldings that have assembled together our contemporary relations with ourselves, I have also tried to argue that they operate according to a common 'diagram'. By 'diagram' I refer to what Deleuze and Guattari describe as an 'abstract machines' – not something that is the cause or origin of all the actual machines we have been investigating, but as immanent within them. An abstract machine in this context is nothing more than a diagram of commonalities, a kind of irreal plane of projection of all the heterogeneous machinations and assemblages – in the way in which, in Foucault's analysis, 'discipline' was the name of a kind of abstract machine that was immanent in the prison, the school, the barracks (Deleuze and Guattari, 1988, pp. 66–7; cf. Foucault, 1977). This diagram, this historical a priori is the positivity opened out by our contemporary regimes of subjectification, a positivity brought into existence by the knowledge and practices of the human sciences, while at the same time establishing for them the very empire they would map, colonize, populate, and connect together by networks of thought and actions. If I might paraphrase Michel Foucault, this 'diagrams' a being that, from the interior of the discourses with which it is surrounded and the practices in which it is assembled, is enabled to know, or required to know, what it is in its positivity – a being that thinks itself both free and determined by positivities essential to itself, that delimits the possibility of its practices of freedom in the very moment in which it accords those positivities the status of truth (cf. Foucault, 1970, p. 353).

This psychological being is now placed at the origin of all the activities of loving, desiring, speaking, laboring, sickening, and dying: the interiority that has been given to humans by all those projects which would seek to know them and act upon them in order to tell them their truth and make possible their improvement and their happiness. It is this being, whose invention is so recent yet so fundamental to our contemporary experience that we today seek to govern under the regulative ideal of freedom – an ideal that imposes as many burdens, anxieties, and divisions as it inspires projects of emancipation, and in the name of which we have come to authorize so many authorities to assist us in the project of being free from any authority but our own. Although we are, no doubt, neither at the dawn of a new age nor at the ending of an old one, we can, perhaps, begin to discern the cracking of this once secure space of interiority, the disconnecting of some of the lines that have made up this diagram, the possibility that, if we cannot disinvent ourselves, we might at least enhance the contestability of the forms of being that have been invented for us, and begin to invent ourselves differently.

Notes

Chapter 1

1 To avoid any confusion, can I point out that subjectification is not used here to imply domination by others, or subordination to an alien system of powers? It functions here not as term of 'critique' but as a device for critical thought – simply to designate processes of being 'made up' as a subject of a certain type. As will be evident, my argument throughout this chapter is dependent upon Michel Foucault's analyses of subjectification.

2 I allude here to a phrase of Michel Maffesoli: "at the heart of the real, then, there is an 'irreal' which is irreducible, and whose action is far from negligible" (Maffesoli, 1991, p. 12).

3 It is important to understand this in the *reflexive,* rather than the substantive mode. In what follows, the phrase always designates this relation, and implies no substantive 'self' as the object of that relation.

4 Of course, this is to overstate the case. One needs to look, on the one hand, at the ways in which philosophical reflections have themselves been organized around problems of pathology – think of the functioning of the image of the statue deprived of all sensory inputs in sensationalist philosophers such as Condillac – and also of the ways in which philosophy is animated by and articulated with, problems of the government of conduct (on Condillac, see Rose, 1985a; on Locke, see Tully, 1993; on Kant, see Hunter, 1994).

5 Similar arguments about the necessity for analyzing 'the self' as technological have been made in a number of quarters recently. See especially the discussion in Elspeth Probyn's recent book (1993). Precisely what is meant by 'technological' is often less clear. As I suggest later and in Chapter 8, an analysis of the technological forms of subjectification needs to develop in terms of the relation between technologies for the government of conduct and the intellectual, corporeal, and ethical techniques that structure being's relation to itself at different moments and sites.

6 This is not, of course, to suggest that knowledge and expertise do not play a crucial role in nonliberal regimes for the government of conduct – one only has to think of the role of doctors and administrators in the organization of the mass extermination programs in Nazi Germany, or of the role of party workers in the pastoral relations of East European states prior to their 'democratization', or the role of planning expertise in centralized planning regimes such as GOSPLAN in the USSR. However, the relations between forms of knowledge and practice designated political and those claiming a nonpolitical grasp of their objects were different in each case.

7 This is not the place to argue this point, so let me just assert that only rationalists, or believers in god, imagine that 'reality' exists in the discursive forms available to thought. This is not a question to be addressed by reviving the old debates over the distinction between knowledge of the 'natural' and 'social' worlds – it is merely to accept that this must be the case unless one believes in some transcendental power that has so shaped human thought such that it is homologous with that which it thinks of. Nor is it to rehearse the old problem of epistemology, which poses an ineffable divide between thought and its object and then perplexes itself as to how one can 'represent' the other. Rather, perhaps one might say that thought makes up the real, but not as a 'realization' of thought.

Chapter 7

1 The argument in this chapter was initially presented at a conference on "The Values of the Enterprise Culture" held at Lancaster University in 1989, and devoted to the analysis of the cultural shifts supposedly brought about by the Conservative regime of Prime Minister Margaret Thatcher, which had at that point been in power for ten years. In republishing it here, under its original title, I have thought it best to keep editorial changes to a minimum of stylistic correction: it should, therefore, be read as an attempt to analyze "what is going on now" when the 'now' in question is the end of the 1980s.

Chapter 8

1 In developing the argument of this chapter, and in particular in utilizing the work of Deleuze and Guattari, I have been greatly helped by reading Elizabeth Grosz's extended meditation on the analytics of bodies (1994). Although I find myself in disagreement with some of her conclusions, my thinking here owes much to her illuminating discussions. Deleuze and Guattari's work has also been taken up in a spate of recently published studies which I have not been able to consider here. Anyone familiar with Deleuze's work will recognize immediately that I have chosen to misunderstand some of his concepts and to eschew many others: for example, the reader will find no 'bodies without organs' and an empiricist reduction of the problematics of desire.

2 I should emphasize again here, as I have elsewhere in this book, that to assert that subjectivity is technological is not to align oneself with the animadversions on the malign effects of the technological order on subjectivity most closely associated with writers of the Frankfurt School. Technology does not crush subjectivity – it produces the possibility of humans relating to themselves as subjects of certain types, as well as the possibilities of their resisting or refusing certain regimes of subjectification.

3 As I was putting this chapter into its final form I came across Constantin Boundas and Dorothea Olkowski's edited collection of essays on Deleuze (1994) and have benefited in particular from the chapter by Boundas (1994).

4 I am reminded here in particular of Donna Haraway's ways of connecting up the enterprise of primatology with the writing of science fiction and its imagining of other forms of relations between creatures (1989, esp. chap. 16).

5 The reference to rhetoric here should indicate that we should not place speech on the side of the natural either.

6 I have benefited here from reading a chapter from Celia Lury's soon to be published study of memory and identity, and would like to thank her for letting me see this in draft form.

7 See Deleuze and Guattari (1994) for some suggestive remarks on 'fleshism'.

8 Of course, many of the writers who emphasize the significance of 'the body' also try to recognize this: that is what seems to be implied by Braidotti's statement that 'the body' is "to be understood as neither a biological nor a sociological category, but rather as a point of overlap between the physical, the symbolic, and the material social conditions" (1984b, p. 161).

9 Bordo quotes this from an article entitled "Exercises for Men" by the Williamette Bridge Liberation News Service, in *The Radical Therapist,* Dec.–Jan., 171.

10 I have adapted Deleuze's language to suit my own purposes here. Foucault's fourfold division – which can itself undoubtedly be traced back to Aristotle – was ontology, ascetics, deontology, and teleology. See Foucault, 1985, 1986a, 1986b; Rose, 1995a; Dean 1994.

Bibliography

Abrams, P. (1968). *The Origins of British Sociology, 1834–1914.* Chicago: University of Chicago Press.

Ahrenfeldt, R. H. (1958). *Psychiatry in the British Army in the Second World War.* London: Routledge and Kegan Paul.

Albig, W. (1956). *Modern Public Opinion.* New York: McGraw-Hill.

Allport, F. H. (1924). *Social Psychology.* Boston: Houghton Mifflin.

(1937). Towards a science of public opinion. *Public Opinion Quarterly,* 1, 7–23.

Allport F. H., and Lepkin, M. (1943). Building war morale with news headlines. *Public Opinion Quarterly,* 7, 211–21.

Allport, F. H., Lepkin, M., and Cahen, E. (1943). Headlines on allied losses are better morale builders. *Editor and Publisher,* 9 October.

Allport, G. W. (1935). Attitudes. In C. A. Murchison, ed., *Handbook of Social Psychology* (pp. 1–50). Worcester, MA: Clark University Press.

(1954). The historical background of modern social psychology. In G. Lindzey, ed., *Handbook of Social Psychology* (pp. 1–80). Cambridge, MA: Addison Wesley.

Allport, G. W., and Postman, L. (1947). *The Psychology of Rumour.* New York: Holt.

Armistead, N., ed. (1974). *Reconstructing Social Psychology.* London: Penguin.

Ash, M. (1995). *Gestalt Psychology in German Culture, 1860–1967: Holism and the Quest for Objectivity.* Cambridge: Cambridge University Press.

Ash, M., and Woodward, W., eds. (1987). *Psychology in Twentieth-Century Thought and Society.* Cambridge: Cambridge University Press.

Bachelard, G. (1984). *The New Scientific Spirit,* tr. Arthur Goldhammer. Boston: Beacon Press (originally published 1934).

Badinter, E. (1981). *The Myth of Motherhood,* tr. R. DeGaris. London: Souvenir Press.

Baistow, K. (1995). Liberation and regulation? Some paradoxes of empowerment. *Critical Social Policy,* 42, 34–46.

Barber, R. (1941). The civilian population under bombardment. *Nature,* 7 June, 700–1.

Baritz, L. (1960). *The Servants of Power: A History of the Use of Social Science in American Industry.* Middletown, CT: Wesleyan University Press.

Barnes, B. (1974). *Scientific Knowledge and Sociological Theory.* London: Routledge and Kegan Paul.

Barnett, A. (1989). "Charlie's army." *New Statesman and Society,* 22 September, 9–11.

Bauer, R. A. (1952). *The New Man in Soviet Psychology.* Cambridge: Cambridge University Press.

Bauman, Z. (1991). *Modernity and Ambivalence.* Oxford: Polity.

Baumeister, R. (1987). How the self became a problem: A psychological review of historical research. *Journal of Personality and Social Psychology,* 52, 1, 163–76.

Bavelas, A., and Lewin, K. (1942). Training in democratic leadership. *Journal of Abnormal and Social Psychology,* 37, 115–19.

Beck, U. (1992). *Risk Society: Towards a New Modernity.* London: Sage.

Bell, D. (1979). *The Cultural Contradictions of Capitalism,* 2nd ed. New York: Basic Books.

Benveniste, E. (1971). *Problems in General Linguistics.* Miami: University of Miami Press.

Bion, W. R. (1961). *Experiences in Groups.* London: Tavistock.

Blacker, C. P. (1946). *Neurosis and the Mental Health Services.* London: Cumberledge.

Bloch, S., and Redaway, P. (1977). *Russia's Political Hospitals: The Abuse of Psychiatry in the Soviet Union.* London: Gollancz.

Bordo, S. (1989). The body and the reproduction of femininity: A feminist appropriation. In A. M. Jaggar and S. R. Bordo, eds., *Gender/Body/Knowledge* (pp. 13–33). New Brunswick, NJ: Rutgers University Press.

(1993). *Unbearable Weight: Feminism, Western Culture and the Body.* Berkeley: University of California Press.

Boring, E. (1929). *A History of Experimental Psychology.* London: Century.

Boundas, C. (1994). Deleuze: Serialization and subject formation. In C. V. Boundas and D. Olkowski, eds., *Gilles Deleuze and the Theater of Philosophy* (pp. 99–118). New York: Routledge.

Boundas, C., and Olkowski, D., eds. (1994). *Gilles Deleuze and the Theater of Philosophy.* New York: Routledge.

Bourdieu, P. (1977). *Outline of a Theory of Practice,* tr. R. Nice. Cambridge: Cambridge University Press.

(1984). *Distinction: A Social Critique of the Judgment of Taste.* London: Routledge and Kegan Paul.

Bowlby, R. (1985). *Just Looking: Consumer Culture in Dreiser, Gissing and Zola.* New York: Methuen.

Box, K., and Thomas, G. (1944). The Wartime Social Survey. *Journal of the Royal Statistical Society,* 107, 151–77.

Braidotti, R. (1994a). *Nomadic Subjects: Embodiment and Sexual Difference in Contemporary Feminist Theory.* New York: Columbia University Press.

(1994b). Towards a new nomadism: Feminist Deleuzian tracks, or metaphysics and metabolism. In C. V. Boundas and D. Olkowski, eds., *Gilles Deleuze and the Theater of Philosophy* (pp. 157–86). New York: Routledge.

Braudel, F. (1985). *Civilization and Capitalism,* vol. 2. London: Fontana.

Bremmer, J., and Roodenburg, H., eds. (1991). *A Cultural History of Gesture.* Cambridge: Polity.

Brown, B., and Cousins, M. (1980). The linguistic fault: The case of Foucault's archaeology. *Economy and Society,* 9, 251–78.

Brown, J. A. C. (1954). *The Social Psychology of Industry.* Harmondsworth: Penguin.

Brown, P. (1989). *The Body and Society.* London: Faber.

Brown, R. K. (1967). Research and consultancy in industrial enterprises. *Sociology,* 1, 33–60.

Brown, W., and Jaques, E. (1965). *Glacier Project Papers.* London: Heinemann.

Bruner, J. (1944). *Mandate from the People.* New York: Duell, Sloan and Pearce.

Bryce, J. (1914). *The American Commonwealth.* New York: Macmillan.

Buck, P. (1985). The social sciences go to war. In Merrit Roe Smith, ed., *Military Enterprise and Technological Change* (pp. 203–52), Cambridge, MA: MIT Press.

Burchell, G. (1991). Peculiar interests: Civil society and "governing the system of natural liberty." In G. Burchell, C. Gordon, and P. Miller, eds., *The Foucault Effect: Studies in Governmentality* (pp. 119–50). Hemel Hempstead: Harvester Wheatsheaf.

Burchell, G., Gordon, C., and Miller, P., eds. (1991). *The Foucault Effect: Studies in Governmentality.* Hemel Hempstead: Harvester Wheatsheaf.

Burckhardt, J. (1990). *The Civilization of the Renaissance in Italy,* tr. S. G. C. Middlemore. London: Penguin (originally published 1860).

Burman, E. (1994). *Deconstructing Developmental Psychology.* London: Routledge.

Burman, E., and Parker, I., eds. (1994). *Discourse Analytic Research.* London: Routledge.

Burns Morton, F. J. (1951). *Foremanship – A Textbook.* London: Chapman and Hall.

Buss, A. R., ed. (1979). *Psychology in Social Context.* New York: Irvington.

Butler, J. (1990). *Gender Trouble: Feminism and the Subversion of Identity.* London: Routledge.

(1993). *Bodies That Matter: On the Discursive Limits of 'Sex'.* London: Routledge.

(1995). For a careful reading. In S. Benhabib, J. Butler, D. Cornell, and N. Fraser, eds., *Feminist Contentions: A Philosophical Exchange* (pp. 127–43). New York: Routledge.

Bynum, W. F., Porter, R., and Shepherd, M., eds. (1985). *The Anatomy of Madness,* 2 vols. London: Tavistock.

(1989). *The Anatomy of Madness,* vol. 3. London: Tavistock.

Callon, M. (1986). Some elements of a sociology of translation. In J. Law, ed., *Power, Action and Belief* (pp. 196–229). London: Routledge and Kegan Paul.

Callon, M., and Latour, B. (1981). Unscrewing the big Leviathan: How actors macrostructure reality and how sociologists help them to do so. In K. Knorr Cetina and A. Cicourel, eds., *Advances in Social Theory* (pp. 277–303). London: Routledge and Kegan Paul.

Canguilhem, G. (1968). *Etudes d'histoire et de philosophie des sciences.* Paris: Vrin.

(1977). *Ideologie et rationalité.* Paris: Vrin.

(1978). *On the Normal and the Pathological.* Dordrecht: Reidel.

Carson, D., ed. (1990). *Risk-Taking in Mental Disorder.* Chichester: SLE Publications.

Cartwright, D. (1947–8). Social psychology in the United States during the Second World War. *Human Relations,* 1, 333–52.

Cartwright, D., and Zander, A. (1953). *Group Dynamics: Research and Theory.* London: Tavistock.

(1967). *Group Dynamics: Research and Theory,* 3rd ed. London: Tavistock.

Castel, R. (1991). From dangerousness to risk. In G. Burchell, C. Gordon, and P. Miller, eds., *The Foucault Effect: Studies in Governmentality* (pp. 281–98). Hemel Hempstead: Harvester Wheatsheaf.

Chartier, R., ed. (1989). *A History of Private Life. Vol. 3: Passions of the Renaissance,* tr. A. Goldhammer. Cambridge, MA: Belknap Press of Harvard University Press.

Child, J. (1969). *British Management Thought.* London: Allen and Unwin.

Cina, C. (1976). Social science research: A tool of counter-insurgency. *Science for the People,* 8, 2, 36.

Clarke, V. M. (1950). *New Times, New Methods and New Men.* London: Allen and Unwin.

Cline-Cohen, P. (1982). *A Calculating People: The Spread of Numeracy in Early America.* Chicago: University of Chicago Press.

Coch, L., and French, J. (1948). Overcoming resistance to change. *Human Relations,* 11, 512–32.

Cocks, G. (1985). *Psychotherapy in the Third Reich: The Göring Institute.* New York: Oxford University Press.

Cohen, A. P. (1994). *Self-Consciousness: An Alternative Anthropology of Identity.* London: Routledge.

Cohen, S., and Scull, A., eds. (1983). *Social Control and the State.* Oxford: Blackwell.

Cole, M., and Maltzman, I., eds., (1969). *A Handbook of Contemporary Soviet Psychology.* New York: Basic Books.

Collini, S. (1979). *Liberalism and Sociology: L. T. Hobhouse and Political Argument in England, 1880–1914.* Cambridge: Cambridge University Press.

(1991). *Public Moralists: Political Thought and Intellectual Life in Britain, 1850–1930.* Oxford: Oxford University Press.

Coward, R., and Ellis, F. (1977). *Language and Materialism: Developments in Semiology and the Theory of the Subject.* London: Routledge and Kegan Paul.

Cruikshank, B. (1993). Revolutions within: Self-government and self-esteem. *Economy and Society,* 22, 3, 326–44.

(1994). The will to empower: Technologies of citizenship and the war on poverty. *Socialist Review,* 23, 4, 29–55.

Cullen, M. (1975). *The Statistical Movement in Early Victorian Britain.* Hassocks, Sussex: Harvester.

Danziger, K. (1988). A question of identity. In J. Morawski, ed., *The Rise of Experimentation in American Psychology* (pp. 35–52). New Haven: Yale University Press.

(1990). *Constructing the Subject: Historical Origins of Psychological Research.* Cambridge: Cambridge University Press.

Dean, M. (1994). "A social structure of many souls": Moral regulation, government and self-formation. *Canadian Journal of Sociology,* 19, 2, 145–68.

(1995). Governing the unemployed self in an active society. *Economy and Society,* 24, 4, 559–83.

Deleuze, G. (1983). *Nietzsche and Philosophy,* tr. Hugh Tomlinson. London: Athlone.

(1988). *Foucault,* tr. S. Hand. Minneapolis: University of Minnesota Press.

(1990a). *Pourparlers.* Paris: Editions de Minuit.

(1990b). *The Logic of Sense,* tr. M. Lester with C. Stivale. New York: Columbia University Press.

(1992a). *The Fold: Leibniz and the Baroque.* Minneapolis: University of Minnesota Press.

(1992b). *Expressionism in Philosophy: Spinoza,* tr. M. Joughin. New York: Zone Books.

Deleuze, G., and Guattari, F. (1988). *A Thousand Plateaus,* tr. B. Massumi. London: Athlone.

(1994). *What Is Philosophy?,* tr. G. Burchell and H. Tomlinson. London: Verso.

Dodds, E. A. (1973). *The Greeks and the Irrational.* Berkeley: University of California Press.

Donnelly, M. (1983). *Managing the Mind.* London: Tavistock.

Donzelot, J. (1979). *The Policing of Families,* with a foreword by G. Deleuze. London: Hutchinson.

Duby, G., ed. (1988). *A History of Private Life. Vol. 2: Revelations of the Medieval World,* tr. A. Goldhammer. Cambridge, MA: Belknap Press of Harvard University Press.

Duden, B. (1991). *The Woman beneath the Skin,* tr. T. Dunlap. Cambridge, MA: Harvard University Press.

du Gay, P. (1995). Making up managers. In S. Hall and P. du Gay, eds., *Questions of Cultural Identity.* London: Sage.

Eisenstein, E. (1979). *The Printing Press as an Agent of Social Change.* Cambridge: Cambridge University Press.

Elias, N. (1978). *The Civilizing Process. Vol. 1: The History of Manners,* tr. Edmund Jephcott. Oxford: Blackwell.

(1983). *The Court Society,* tr. E. Jephcott. Oxford: Blackwell.

Eräsaari, L. (1991). Bureaucratic Space. In J. Lehto, ed., *Deprivation, Social Welfare and Expertise* (pp. 167–79). Helsinki: National Agency for Welfare and Health.

Ewald, F. (1991). Insurance and risk. In G. Burchell, C. Gordon, and P. Miller, eds., *The Foucault Effect: Studies in Governmentality* (pp. 197–210). Hemel Hempstead: Harvester Wheatsheaf.

Ewen, S. (1976). *Captains of Consciousness.* New York: Basic Books.

(1988). *All Consuming Images: The Politics of Style in Contemporary Culture.* New York: Basic Books.

Farago, L., and Gittler, L. F., eds. (1942). *German Psychological Warfare.* New York: Putnam.

Farmer, E. (1958). Early days in industrial psychology: An autobiographical note. *Occupational Psychology,* 32, 264–7.

Farr, R. M. (1978). On the varieties of social psychology: An essay on the relationships between psychology and other social sciences. *Social Science Information,* 17, 503–25.

Feher, M., with Nadaff, R., and Tazi, N., eds. (1989a). *Fragments for a History of the Human Body, Part One.* New York: Zone.

(1989b). *Fragments for a History of the Human Body, Part Two.* New York: Zone.

(1989c). *Fragments for a History of the Human Body, Part Three.* New York: Zone.

Fisher, V. E., and Hanna, J. V. (1932). *The Dissatisfied Worker.* New York: Macmillan.

Forquet, F. (1980). *Les comptes de la puissance.* Encres: Editions Recherches.

Foucault, M. (1967). *Madness and Civilization: A History of Insanity in the Age of Reason.* London: Tavistock.

(1969). Qu'est-ce qu'un auteur? *Bulletin de la Sociéte Française de Philosophie,* 63.

Translated as What Is an Author? in P. Rabinow, ed. 1984, *The Foucault Reader* (pp. 101–20). Harmondsworth: Penguin.
(1970). *The Order of Things.* London: Tavistock.
(1972a). *The Archaeology of Knowledge.* London: Tavistock.
(1972b). Orders of Discourse. *Social Science Information,* 10, 7–30.
(1973). *Birth of the Clinic: An Archaeology of Medical Perception.* London: Tavistock.
(1977). *Discipline and Punish: The Birth of the Prison.* London: Allen Lane.
ed. (1978a). *I, Pierre Riviere, Having Slaughtered My Mother, My Sister and My Brother..* Harmondsworth: Penguin.
(1978b). Politics and the study of discourse. *Ideology and Consciousness,* 3, 7–26.
(1979a). On governmentality. *I & C,* 6, 5–21.
(1979b). *The History of Sexuality. Vol. 1: An Introduction.* London: Allen Lane.
(1981). Omnes et singulatim: Towards a criticism of "political reason." In S. McMurrin, ed., *The Tanner Lectures on Human Values II* (pp. 225–54). Salt Lake City: University of Utah Press.
(1982). The subject and power. Afterword to H. Dreyfus and P. Rabinow, eds., *Michel Foucault: Beyond Structuralism and Hermeneutics* (pp. 208–27). Chicago: University of Chicago Press.
(1985). *The Use of Pleasure.* London: Penguin Viking.
(1986a). On the genealogy of ethics: An overview of work in progress. In P. Rabinow, ed., *The Foucault Reader* (pp. 340–72). Harmondsworth: Penguin.
(1986b). *The History of Sexuality. Vol. 3: The Care of the Self,* tr. R. Hurley. New York: Pantheon.
(1986c). "What is Enlightenment?" *Economy and Society,* 15, 1, 88–96.
(1986d). Space, Knowledge and Power. In P. Rabinow, ed., *The Foucault Reader* (pp. 239–56). Harmondsworth: Penguin.
(1988). Technologies of the self. In L. H. Martin, H. Gutman, and P. H. Hutton, eds., *Technologies of the Self* (pp. 16–49). London: Tavistock.
(1991). Governmentality. In G. Burchell, C. Gordon, and P. Miller, eds. *The Foucault Effect: Studies in Governmental Rationality* (pp. 87–104). Hemel Hempstead: Harvester Wheatsheaf.
Fraser, N. (1989). Foucault on modern power: Empirical insights and normative confusions. In *Unruly Practices* (pp. 17–34). Minneapolis: University of Minnesota Press.
Freidson, E. (1970). *Profession of Medicine.* New York: Harper and Row.
(1986). *Professional Powers: A Study of the Institutionalization of Formal Knowledge.* Chicago: University of Chicago Press.
Freud, S. (1953–7). *Studies in Hysteria.* In J. Strachey, ed., *Standard Edition of the Collected Works of Sigmund Freud,* vol. 2. London: Hogarth Press.
Friedman, M. (1982). *Capitalism and Freedom.* Chicago: University of Chicago Press.
Gallup, G., and Rae, S. F. (1940). *The Pulse of Democracy: The Public Opinion Poll and How It Works.* New York: Simon and Schuster.
Garland, D. (1985). *Punishment and Welfare.* London: Gower.
Gatens, M. (1991). *Feminism and Philosophy: Perspectives on Difference and Equality.* Cambridge: Polity.
Geertz, C. (1979). From the native's point of view: On the nature of anthropological understanding. In P. Rabinow and W. M. Sullivan, eds., *Interpretive Social Science* (pp. 225–42). Berkeley: University of California Press.

Gergen, K. J. (1985a). The social constructionist movement in modern psychology. *American Psychologist,* 40, 266–75.

(1985b). Social psychology and the phoenix of unreality. In S. Koch and D. Leary, eds., *A Century of Psychology as a Science* (pp. 528–57). New York: McGraw-Hill.

(1985c). Social constructionist inquiry: Context and implications. In K. J. Gergen and K. E. Davis, eds., *The Social Construction of the Person* (pp. 3–18). New York: Springer Verlag.

Gergen, K. J., and Davis, K. E., eds. (1985). *The Social Construction of the Person.* New York: Springer Verlag.

Gergen, K. J., and Gergen, M. (1988). Narrative and the self as relationship. In L. Berkowitz, ed., *Advances in Experimental and Social Psychology,* vol. 21 (pp. 17–56). New York: Academic Press.

Geuter, U. (1992). *The Professionalization of Psychology in Nazi Germany.* Cambridge: Cambridge University Press.

Giddens, A. (1991). *Modernity and Self-Identity: Self and Society in the Late Modern Age.* Cambridge: Polity.

Gigerenzer, G. (1991). From tools to theories: A heuristic of discovery in cognitive psychology. *Psychological Review,* 98, 254–67.

Gilman, S. (1982). *Seeing the Insane.* New York: Wiley.

(1985). *Difference and Pathology: Stereotypes of Sexuality, Race and Madness.* Ithaca, NY: Cornell University Press.

Ginneken, J. van (1992). *Crowds, Psychology and Politics, 1871–1899.* Cambridge: Cambridge University Press.

Glover, E. (1940). The birth of social psychiatry. *Lancet,* 24 August, 239.

Goody, J. (1977). *The Domestication of the Savage Mind.* Cambridge: Cambridge University Press.

Goody, J., and Watt, I. (1968). The consequences of literacy. In J. Goody, ed., *Literacy in Traditional Societies* (pp. 27–84). Cambridge: Cambridge University Press.

Gordon, C. (1980). Afterword. In C. Gordon, ed., *Michel Foucault: Power/Knowledge* (pp. 229–59). Brighton: Harvester.

(1986). Question, ethos, event: Foucault on Kant and enlightenment. *Economy and Society,* 15, 71–87.

(1987). The soul of the citizen: Max Weber and Michel Foucault on rationality and government. In S. Lash and S. Whimster, eds., *Max Weber, rationality and modernity* (pp. 293–316). London: Allen and Unwin.

(1991). Governmental rationality: An introduction. In G. Burchell, C. Gordon, and P. Miller, eds., *The Foucault Effect: Studies in Governmental Rationality* (pp. 1–51). Hemel Hempstead: Harvester.

Gordon, L. (1989). *Heroes of Their Own Lives: The Politics and History of Family Violence.* London: Virago.

Grey, S. G. (1975). The Tavistock Institute of Human Relations. In H. V. Dicks, ed., *Fifty Years of the Tavistock Clinic* (pp. 206–27). London: Routledge and Kegan Paul.

Grosz, E. (1993). Bodies and knowledges: Feminism and the crisis of reason. In A. Alcoff and E. Potter, eds., *Feminist Epistemologies.* London: Routledge.

(1994). *Volatile Bodies: Toward a Corporeal Feminism.* St. Leonards, Australia: Allen and Unwin.

Habermas, J. (1971). *Toward a Rational Society.* London: Heinemann.

(1972). *Knowledge and Human Interests.* London: Heinemann.

(1984). *The Theory of Communicative Action. Vol. 1: Reason and the Rationalization of Society.* London: Heinemann.

(1987). The normative content of modernity. In *The Philosophical Discourse of Modernity* (pp. 336–67). Cambridge: Polity.

Hacking, I. (1981). How should we do the history of statistics? *I & C,* 8, 15–26.

(1986). Making up people. In T. C. Heller, M. Sosna, and D. E. Wellberg, eds., *Reconstructing Individualism* (pp. 222–36). Stanford, CA: Stanford University Press.

(1990). *The Taming of Chance.* Cambridge: Cambridge University Press.

Hadot, P. (1992). Reflections on the notion of 'the cultivation of the self'. In T. J. Armstrong, ed., *Michel Foucault, Philosopher* (pp. 225–32). Hemel Hempstead: Harvester Wheatsheaf.

Haraway, D. (1989). *Primate Visions: Gender, Race, and Nature in the World of Modern Science.* New York: Routledge.

(1991). A cyborg manifesto: Science, technology and socialist feminism in the late twentieth century. In *Simians, Cyborgs and Women: The Re-Invention of Nature* (pp. 149–81). New York: Routledge.

Harré, R. (1983). *Personal Being.* Oxford: Blackwell.

(1985). The language game of self ascription: A note. In K. J. Gergen and K. E. Davis, eds., *The Social Construction of the Person* (pp. 259–64). New York: Springer Verlag.

(1989). Language games and texts of identity. In J. Shotter and K. Gergen, eds., *Texts of Identity* (pp. 20–35). London: Sage.

Hayek, F. A. (1976). *The Constitution of Liberty.* London: Routledge and Kegan Paul.

Heelas, P., and Lock, A. (1981). *Indigenous Psychologies: The Anthropology of the Self.* London: Academic Press.

Hennis, W. (1987). Max Weber's theme: Personality and life orders. In S. Whimster and S. Lash, eds., *Max Weber, Rationality and Modernity* (pp. 52–74). London: Allen and Unwin.

Hermans, H., and Kempen, H. (1993). *The Dialogic Self: Meaning as Movement.* New York: Academic Press.

Hirst, P., and Woolley, P. (1982). *Social Relations and Human Attributes.* London: Tavistock.

Horvath, A., and Szakolczai, A. (1992). *The Dissolution of Communist Power: The Case of Hungary.* London: Routledge.

Hovland, C., Janis, I., and Kelley, T. (1953). *Communication and Persuasion: Psychological Studies of Opinion Change.* New Haven: Yale University Press.

Humphreys, P. (1985). *Social Psychology: Development, Experience and Behaviour in a Social World.* Milton Keynes: Open University Press.

Hunter, I. (1988). *Culture and Government: The Emergence of Literary Education.* London: Macmillan.

(1993a). Culture, bureaucracy and the history of popular education. In D. Meredyth and D. Tyler, eds., *Child and Citizen: Genealogies of Schooling and Subjectivity* (pp. 11–34). Queensland: Griffith University, Institute of Cultural Policy Studies.

(1993b). The pastoral bureaucracy: Towards a less principled understanding of state schooling. In D. Meredyth and D. Tyler, eds., *Child and Citizen: Genealogies of Schooling and Subjectivity* (pp. 237–87). Queensland: Griffith University, Institute of Cultural Policy Studies.

(1993c). Subjectivity and government. *Economy and Society,* 22, 1, 123–34.

(1994). *Rethinking the School: Subjectivity, Bureaucracy, Criticism.* St. Leonards, Australia: Allen and Unwin.

Irigaray, L. (1985). *Speculum of the Other Woman,* tr. G. Gill. Ithaca, NY: Cornell University Press.

Jambet, C. (1992). The constitution of the subject and spiritual practice. In T. J. Armstrong, Ed., *Michel Foucault, Philosopher* (pp. 233–47). Hemel Hempstead: Harvester Wheatsheaf.

Janis, I. (1951). *Air War and Emotional Stress.* New York: McGraw-Hill.

Jaques, E. (1951). *The Changing Culture of a Factory.* London: Tavistock.

Jones, K., and Williamson, K. (1979). The birth of the schoolroom. *I & C,* 6, 59–110.

Jones, M. (1952). *Social Psychiatry.* London: Tavistock.

Joravsky, D. (1989). *Russian Psychology.* Oxford: Blackwell.

Joyce, P. (1994). *Democratic Subjects: The Self and the Social in Nineteenth Century England.* Cambridge: Cambridge University Press.

Katz, D. (1946). The Surveys Division of OWI: Governmental use of research for informational purposes. In A. B. Blankenship, ed., *How to Conduct Consumer and Opinion Research.* New York: Harper.

Kozulin, A. (1984). *Psychology in Utopia: Toward a Social History of Soviet Psychology.* Cambridge, MA: MIT Press.

Kraupl-Taylor, F. (1958). A history of group and administrative therapy in Great Britain. *British Journal of Medical Psychology,* 31, 153–73.

Krech, D., Crutchfield, R., and Ballachy, E. L. (1962). *Individual in Society.* New York: McGraw-Hill.

Lacan, J. (1977). The agency of the letter in the unconscious or reason since Freud. In J. Lacan, *Écrits,* tr. A. Sheridan. London: Tavistock.

Laqueur, T. (1990). *Making Sex: Body and Gender from the Greeks to Freud.* Cambridge, MA: Harvard University Press.

Lasch, C. (1979). *The Culture of Narcissism.* New York: Norton.

(1984). *The Minimal Self: Psychic Survival in Troubled Times.* New York: Norton.

Lash, S., and Friedman, J., eds. (1992). *Modernity and Identity.* Oxford: Blackwell.

Lash, S., Heelas, P., and Morris, P., eds. (1995). *De-Traditionalization: Authority and Self in an Age of Cultural Uncertainty.* Oxford: Blackwell.

Lasswell, H. D. (1926). *Propaganda Technique in the World War.* New York: Knopf.

(1941). *Democracy through Public Opinion.* Menasha, WI: Banta.

Latour, B. (1984). *Les microbes, guerre et paix.* Paris: Metailie. Translated as *The Pasteurisation of French Society* by A. Sheridan and J. Law. Cambridge, MA: Harvard University Press.

(1986a). The powers of association. In J. Law, ed., *Power, Action and Belief* (pp. 264–80). London: Routledge and Kegan Paul.

(1986b). Visualization and cognition: Thinking with eyes and hands. *Knowledge and Society: Studies in the Sociology of Culture Past and Present,* 6, 1–40.

(1988). *Science in Action.* Milton Keynes: Open University Press.

Law, J., ed. (1986). *Power, Action and Belief.* London: Routledge and Kegan Paul.

(1987). Technology and heterogeneous engineering: The case of Portuguese expansion. In W. Bijker, T. P. Hughes, and T. Pinch, eds., *The Social Construction of Technological Systems* (pp. 111–34). Cambridge, MA: MIT Press.

Lazarsfeld, P., Gaudet, H., and Berelson, B. (1945). *The People's Choice.* New York: Duell, Sloan and Pearce.

Le Bon, G. (1895). *The Crowd: A Study of the Popular Mind.* London: Fisher Unwin.

Leighton, A. H. (1945). *The Governing of Men: General Principles and Recommendations Based on Experience at a Japanese Relocation Camp.* Princeton: Princeton University Press.

(1949). *Human Relations in a Changing World: Observations on the Use of the Social Sciences.* New York: Dutton.

Lerner, D. (1949). *Sykewar: Psychological Warfare against Germany, D-Day to VE-Day.* New York: Stewart.

Lewin, K. (1948). *Resolving Social Conflicts: Selected Papers on Group Dynamics,* ed. G. W. Lewin. New York: Harper and Row.

Lewin, K., Lippitt, R., and White, R. (1939). Patterns of aggressive behavior in experimentally created 'social climates'. *Journal of Social Psychology,* 10, 271–99.

Lewis, A. (1942). Incidence of neurosis in England under war conditions. *Lancet,* 15 August, 175.

Lippitt, R. (1939). Field theory and experiment in social psychology: Autocratic and democratic group atmospheres. *American Journal of Sociology,* 45, 26–49.

(1940). Studies on experimentally created autocratic and democratic groups. *University of Iowa Studies: Studies in Child Welfare,* 16, 3, 45–198.

Lippmann, W. (1922). *Public Opinion.* London: Allen and Unwin.

Lloyd, G. (1984). *The Man of Reason: 'Male' and 'Female' in Western Philosophy.* London: Methuen.

London Centre for Psychotherapy (1987). *Untitled Brochure.* London: London Centre for Psychotherapy.

Lowell, A. Lawrence (1926). *Public Opinion and Popular Government.* New York: Longmans Green.

Lynch, M. (1985). Discipline and the material form of images: An analysis of scientific visibility. *Social Studies of Science,* 15, 37–66.

Lynch, M., and Woolgar, S. (1988). Sociological orientations to representational practices in science. Introduction to M. Lynch and S. Woolgar, eds., *Representational Practices in the Natural Sciences.* Special issue of *Human Studies,* 11, 1.

MacDonagh, O. (1958). The nineteenth century revolution in government: A reappraisal. *Historical Journal,* 1, 1, 52–67.

(1977). *Early Victorian Government, 1830–1870.* London: Weidenfeld and Nicholson.

Mace, C. A. (1948). Satisfactions in work. *Occupational Psychology,* 22, 5.

Mace, C. A., and Vernon, P. E., eds. (1953). *Current Trends in British Psychology.* London: Methuen.

MacIntyre, A. (1981). *After Virtue: A Study in Moral Theory.* London: Duckworth.

Mackay, D. (1984). Behavioural psychotherapy. In W. Dryden, ed., *Individual Therapy in Britain* (pp. 264–94). London: Harper and Row.

MacKenzie, D. (1981). *Statistics in Britain, 1865–1930: The Social Construction of Scientific Knowledge.* Edinburgh: Edinburgh University Press.

MacLeod, R. (1988). *Government and Expertise: Specialists, Administrators and Professionals, 1860–1919.* Cambridge: Cambridge University Press.

Madge, C., and Harrison, T., eds. (1938). *Mass Observation: First Year's Work, 1937–8,* with an essay on a Nation Wide Intelligence Service by Bronislaw Malinowski. London: Mass Observation.
Madge, J. (1963). *The Origins of Scientific Sociology.* London: Tavistock.
Maffesoli, M. (1991). The ethics of aesthetics. *Theory, Culture and Society,* 8, 7–20.
Main, T. (1946). The hospital as a therapeutic institution. *Bulletin of the Menninger Clinic,* 10, 66–70.
Markus, T. A. (1993). *Buildings and Power: Freedom and Control in the Origin of Modern Building Types.* London: Routledge.
Marrow, A. J. (1957). *Making Management Human.* New York: McGraw-Hill.
(1969). *The Practical Theorist: The Life and Work of Kurt Lewin.* New York: Basic Books.
Mauss, M. (1979a). Body techniques. In *Psychology and Sociology: Essays* (pp. 95–123). London: Routledge and Kegan Paul.
(1979b). The category of the person. In *Psychology and Sociology: Essays* (pp. 57–94). London: Routledge and Kegan Paul.
Mayo, E. (1933). *The Human Problems of an Industrial Civilization.* New York: Macmillan.
McDougall, W. (1920). *The Group Mind.* Cambridge: Cambridge University Press.
McLaine, I. (1979). *Ministry of Morale: Home Front Morale and the Ministry of Information in World War II.* London: George Allen and Unwin.
Mesmer, F. A. (1957). *Mesmerism: A Translation of the Original Medical and Scientific Writings,* tr. G. J. Bloch. Los Altos, CA: Kaufman.
Meyer, J. (1986). The self and the life course: Institutionalization and its effects. In A. Sørensen, F. Weinert, and L. Sherrod, eds., *Human Development and the Life-Course* (pp. 199–216). Hillsdale, NJ: Erlbaum.
Middlemas, K. (1979). *Politics in Industrial Society.* London: Deutsch.
Miller, P. (1986). Psychotherapy of work and unemployment. In P. Miller and N. Rose, eds., *The Power of Psychiatry* (pp. 143–76). Cambridge: Polity.
(1987). *Domination and Power.* London: Routledge and Kegan Paul.
(1989). Accounting for the calculable self. Paper delivered to the Workshop on Controlling Social Life, European University Institute, Florence, May.
(1992). Accounting and objectivity: The invention of calculating selves and calculable spaces. *Annals of Scholarship,* 9, 1–2, 61–86.
Miller, P., and O'Leary, T. (1987). Accounting and the construction of the governable person. *Accounting, Organizations and Society,* 12, 3, 235–65.
Miller, P., and O'Leary, T. (1989). Hierarchies and American ideals, 1900–1940. *Academy of Management Review,* 14, 2, 250–65.
Miller, P., and Rose, N., eds. (1986). *The Power of Psychiatry.* Cambridge: Polity.
(1988). The Tavistock programme: Governing subjectivity and social life. *Sociology,* 22, 171–92.
(1989). Poliittiset rationalisaatiot ja hallintekniikat. *Polittiikka,* 1989, 3, 145–58. Translated as Political rationalities and technologies of government. In S. Hanninen and K. Palonen, eds., *Texts, Contexts, Concepts* (pp. 171–83). Helsinki: Finnish Political Science Association.
(1990). Governing economic life. *Economy and Society,* 19, 1, 1–31.
(1994). On therapeutic authority. *History of the Human Sciences,* 7, 3, 29–64.
(1995). Production, identity and democracy. *Theory and Society,* 24, 427–67.

(1996). Mobilising the consumer: Assembling the subject of consumption. *Theory, Culture and Society.*

(forthcoming). *In Search of Human Relations: A Social and Intellectual History of the Tavistock Clinic and the Tavistock Institute of Human Relations.* London: Routledge.

Minson, J. P. (1993). *Questions of Conduct.* London: Macmillan.

Morawski, J. G., ed. (1988a). *The Rise of Experimentation in American Psychology.* New Haven: Yale University Press.

(1988b). Impossible experiments and practical constructions. In J. G. Morawski, ed., *The Rise of Experimentation in American Psychology* (pp. 72–93). New Haven: Yale University Press.

Moscovici, S. (1985). *The Age of the Crowd: A Historical Treatise on Mass Psychology,* tr. J. C. Whitehouse. Cambridge: Cambridge University Press.

Musil, R. (1979). *The Man without Qualities,* vol. 1. London: Picador (originally published 1930).

Myers, C. S. (1927). *Industrial Psychology in Great Britain.* London: Cape.

NHS Management Executive (1993). *Risk Management in the NHS.* London: Department of Health.

Nietzsche, F. (1956). *The Genealogy of Morals,* tr. Francis Golffing. New York: Doubleday (originally published 1887).

(1973). *Beyond Good and Evil.* Harmondsworth: Penguin (originally published 1886).

Northcott, C. H. (1945). *Personnel Management: Its Scope and Practice.* London: Pitman.

Oestreich, G. (1982). *Neostoicism and the Modern State.* Cambridge: Cambridge University Press.

Ong, W. (1982). *Orality and Literacy.* London: Methuen.

Osborne, T. (1994). Bureaucracy as a vocation: Liberalism, ethics and administrative expertise in the nineteenth century. *Journal of Historical Sociology,* 7, 3, 289–313.

(1996). Constructionism, authority and the ethical life. In I. Velody and R. Williams, eds., *Social Constructionism.* London: Sage.

Parker, I., and Shotter, J., eds. (1990). *Deconstructing Social Psychology.* London: Routledge.

Parsons, T. (1939). The professions and social structure. *Social Forces,* 17, 4. Reprinted in *Essays in Sociological Theory Pure and Applied.* Glencoe, IL: Free Press, 1949.

(1951a). *The Social System.* Glencoe, IL: Free Press.

(1951b). Illness and the role of the physician. *American Journal of Orthopsychiatry,* 21, 452–60.

Parsons, T., and Shils, E. A., eds. (1951). *Toward a General Theory of Action.* Cambridge, MA: Harvard University Press.

Pasquino, P. (1978). Theatrum Politicum. The genealogy of capital – police and the state of prosperity. *Ideology and Consciousness,* 4, 41–54.

(1991). Criminology: The birth of a special knowledge. In G. Burchell, C. Gordon, and P. Miller, eds., *The Foucault Effect: Studies in Governmentality* (pp. 235–50). Hemel Hempstead: Harvester.

Patton, P. (1994). Foucault's subject of power. *Political Theory Newsletter,* 6, 1, 60–71.

Perrot, M., ed. (1990). *A History of Private Life. Vol. 4: From the Fires of Revolution to the Great War,* tr. A. Goldhammer. Cambridge, MA: Belknap Press of Harvard University Press.

Pick, D. (1989). *Faces of Degeneration: A European Disorder, c. 1848 – c. 1918.* Cambridge: Cambridge University Press.

Plato (1973). *Phaedrus and Letters VII and VIII,* tr. W. Hamilton. Harmondsworth: Penguin.

Potter, J., and Wetherell, M. (1984). *Discourse and Social Psychology.* London: Routledge.

Probyn, E. (1993). *Sexing the Self: Gendered Positions in Cultural Studies.* London: Routledge.

Prost, A., and Vincent, G., eds. (1991). *A History of Private Life. Vol. 5: Riddles of Identity in Modern Times,* tr. A. Goldhammer. Cambridge, MA: Belknap Press of Harvard University Press.

Qualter, T. H. (1985). *Opinion Control in the Democracies.* London: Macmillan.

Rabinow, P., ed. (1984). *The Foucault Reader.* Harmondsworth: Penguin.

Rajchman, J. (1991). *Truth and Eros: Foucault, Lacan and the Question of Ethics.* New York: Routledge.

Ranum, O. (1989). The refuges of intimacy. In R. Chartier, ed., *A History of Private Life. Vol. 3: Passions of the Renaissance,* tr. A. Goldhammer (pp. 207–63). Cambridge, MA: Belknap Press of Harvard University Press.

Rhodes, R. (1987). *The Making of the Atomic Bomb.* New York: Simon and Schuster.

Richards, B. (1989). *Images of Freud.* London: Dent.

Rieff, P. (1959). *Freud: The Mind of the Moralist.* London: Gollancz.

(1966). *The Triumph of the Therapeutic: Uses of Faith after Freud.* Chicago: University of Chicago Press.

(1987). *The Triumph of the Therapeutic: Uses of Faith after Freud,* with a new preface by Philip Rieff. Chicago: University of Chicago Press.

Riley, D. (1983). *War in the Nursery: Theories of the Child and the Mother.* London: Virago.

(1988). *Am I That Name?* London: Macmillan.

Ritchie, J. H., Dick, D., and Lingham, R. (1994). *The Report of the Inquiry into the Care and Treatment of Christopher Clunis.* London: HMSO.

Robinson, C. E. (1937). Recent developments in the straw poll field. *Public Opinion Quarterly,* 1, 45–67.

Roethlisberger, F. J., and Dickson, W. J. (1939). *Management and the Worker.* Cambridge, MA: Harvard University Press.

Rogers, C. R. (1961). *On Becoming a Person.* London: Constable.

Rollins, N. (1972). *Child Psychiatry in the Soviet Union: Preliminary Observations.* Cambridge, MA: Harvard University Press.

Rose, M. (1978). *Industrial Behaviour: Theoretical Developments since Taylor.* Harmondsworth: Penguin.

Rose, N. (1979). The psychological complex: Mental measurement and social administration. *Ideology and Consciousness,* 5, 5–68.

(1984). The administration of children. Unpublished paper delivered to the British Psychological Society (Developmental Psychology Section), Annual Conference, September.

(1985a). *The Psychological Complex: Psychology, Politics and Society in England, 1869–1939.* London: Routledge and Kegan Paul.

(1985b). Michel Foucault and the study of psychology. *Psychcritique,* 1, 133–7.

(1986a). Psychiatry: The discipline of mental health. In P. Miller and N. Rose, eds., *The Power of Psychiatry* (pp. 43–84). Cambridge: Polity.

(1987). Beyond the public/private division: Law, power and the family. *Journal of Law and Society,* 14, 1, 61–76.

(1988). Calculable minds and manageable individuals. *History of the Human Sciences,* 1, 179–200.

(1989a). Social psychology as a science of democracy. In *Proceedings of 8th Cheiron-Europe Conference.* Goteborg: Cheiron.

(1989b). Psychology as a 'social' science. In I. Parker and J. Shotter, *Deconstructing Social Psychology* (pp. 103–16). London: Routledge.

(1990). *Governing the Soul: The Shaping of the Private Self.* London: Routledge.

(1991). Governing by numbers: Figuring out democracy. *Accounting, Organizations and Society,* 16, 7, 673–92.

(1992a). Governing the enterprising self. In P. Heelas and P. Morris, eds., *The Values of the Enterprise Culture: The Moral Debate* (pp. 141–64). London: Routledge.

(1992b). Engineering the human soul: Analysing psychological expertise. *Science in Context,* 5, 2, 351–69.

(1993). *Towards a Critical Sociology of Freedom.* Inaugural Lecture delivered on 5 May 1992 at Goldsmiths' College, University of London. Goldsmiths' College Occasional Paper. London.

(1994a). Government, authority and expertise under advanced liberalism. *Economy and Society,* 22, 3, 273–99 .

(1994b). Expertise and the government of conduct. *Studies in Law, Politics and Society,* 14, 359–97.

(1995a). Authority and the genealogy of subjectivity. In S. Lash, P. Heelas, and P. Morris, eds., *De-Traditionalization: Authority and Self in an Age of Cultural Change.* Oxford: Blackwell.

(1995b). Identity, genealogy, history. In S. Hall and P. du Gay, eds., *Questions of Cultural Identity.* London: Sage.

Rose, N., and Miller, P. (1989). Rethinking the state: Governing economic, social and personal life. Unpublished manuscript.

(1992). Political power beyond the state: Problematics of government. *British Journal of Sociology,* 43, 2, 172–205.

Sampson, E. (1989). The deconstruction of the self. In J. Shotter and K. Gergen, eds., *Texts of Identity* (pp. 1–19). London: Sage.

Schumpeter, J. A. (1954). *History of Economic Analysis.* New York: Oxford University Press.

Scull, A. (1979). *Museums of Madness.* London: Allen Lane.

(1989). From madness to mental illness: Medical men as moral entrepreneurs. In *Social Order/Mental Disorder* (pp. 118–61). London: Routledge.

Sedgwick, E. (1993). Epidemics of the will. In J. Crary and S. Kwinter, eds., *Incorporations* (pp. 582–95). New York: Zone.

Self-Helpline (1989). Who else would you ask if you wanted an answer to these questions [advertisement]? *Guardian,* 18 August, p. 1.

Sennett, R. (1977). *The Fall of Public Man.* London: Faber.

Sherif, M. (1936). *The Psychology of Social Norms.* New York: Octagon.

Sherif, M., and Sherif, C. W. (1953). *Groups in Harmony and Tension.* New York: Harper.

Shields, R., ed. (1992). *Lifestyle Shopping: The Subject of Consumption.* London: Routledge.

Shortland, M. (1985). Barthes, Lavater and the legible body. *Economy and Society,* 14, 4, 273–312.

Shotter, J. (1985). Social accountability and self specification. In K. J. Gergen and K. E. Davies, eds., *The Social Construction of the Person* (pp. 168–90). New York: Springer Verlag.

(1989). Social accountability and the social construction of 'you'. In J. Shotter and K. Gergen, eds., *Texts of Identity* (pp. 133–51). London: Sage.

Shotter, J., and Gergen, K., eds. (1989). *Texts of Identity.* London: Sage.

Showalter, E. (1987). *A Female Malady: Women, Madness and English Culture, 1830–1980.* London: Virago.

Smith, B. L., Lasswell, H. D., and Casey, R. S. (1946). *Propaganda, Communication and Public Opinion.* Princeton: Princeton University Press.

Smith, R. (1981). *Trial by Medicine: Insanity and Responsibility in Victorian Trials.* Edinburgh: Edinburgh University Press.

(1988). Does the history of psychology have a subject? *History of the Human Sciences,* 1, 2, 147–77.

(1992). *Inhibition: History and Meaning in the Sciences of Mind and Brain.* Berkeley: University of California Press.

Sprawson, C. (1992). *Haunts of the Black Masseur: The Swimmer as Hero.* London: Jonathon Cape.

Stouffer, S. A., et al. (1949a). *The American Soldier. Vol. I: Adjustment during Army Life.* New York: Wiley.

(1949b). *The American Soldier. Vol. II: Combat and Its Aftermath.* New York: Wiley.

(1950a). *The American Soldier. Vol. III: Experiments in Mass Communication.* New York: Wiley.

(1950b). *The American Soldier. Vol. IV: Measurement and Prediction.* New York: Wiley.

Strathern, M. (1992). *After Nature.* Cambridge: Cambridge University Press.

Summa, H. (1990). Ethos, pathos and logos in central governmental texts: Rhetoric as an approach to studying a policy making process. In S. Hanninen and K. Palonen, eds., *Texts, Contexts, Concepts* (pp. 184–201). Helsinki: Finnish Political Science Association.

Taussig, M. (1993). *Mimesis and Alterity: A Particular History of the Senses.* New York: Routledge.

Taylor, C. (1989). *Sources of the Self: The Making of Modern Identity.* Cambridge: Cambridge University Press.

Taylor, F. W. (1913). *Principles of Scientific Management.* New York: Harper.

Taylor, G. R. (1950). *Are Workers Human?* London: Falcon Press.

Thomas, W. I., and Znaniecki, F. (1918). *The Polish Peasant in Europe and America. Vol. 1: Primary Group Organization.* Boston: Badger.

Thurstone, L. L. (1928). Attitudes can be measured. *American Journal of Sociology,* 33, 529–54.

Thurstone, L. L., and Chave, E. J. (1929). *The Measurement of Attitudes.* Chicago: University of Chicago Press.

Titmuss, R. (1950). *Problems of Social Policy.* History of the Second World War, UK Civil Series. London: HMSO.

Tredgold, R. F. (1948). Mental hygiene in industry. In N. G. Harris, ed., *Modern Trends in Psychological Medicine.* London: Butterworth.

(1949). *Human Relations in Modern Industry.* London: Duckworth.

Tribe, K. (1976). *Land, Labour and Economic Discourse.* London: Routledge and Kegan Paul.

Tully, J. (1993). Governing conduct. In *An Approach to Political Philosophy: Locke in contexts.* Cambridge: Cambridge University Press.

United States Congress (1973). *Abuse of Psychiatry for Political Repression in the Soviet Union.* New York: Arno Press.

United States Strategic Bombing Survey (1947). *The Effects of Strategic Bomb ing on German Morale,* 2 vols. Washington, DC: U.S. Government Printing Office.

Urwick, L., and Brech, E. F. L. (1948). *The Making of Scientific Management,* vol. 3. London: Management Publications.

Ussher, J. (1991). *Women's Madness.* Hemel Hempstead: Harvester Wheatsheaf.

Valverde, M. (1985). *Sex, Power and Pleasure.* Toronto: Women's Press.

(1991). The Age of Light, Soap, and Water. Toronto: McClelland and Stewart.

(1996). Despotism and ethical liberal-governance. *Economy and Society.*

Vernon, P. E. (1941). A study of war attitudes. *British Journal of Medical Psychology,* 19, 271–91.

Veyne, P., ed. (1987). *A History of Private Life. Vol. 1: From Pagan Rome to Byzantium,* tr. A. Goldhammer. Cambridge, MA: Belknap Press of Harvard University Press.

Vico, G. (1848). *The New Science of Giambattista Vico* (1725), translated from 3rd ed., 1744. Ithaca, NY: Cornell University Press, 1984.

Viteles, M. S. (1932). *Industrial Psychology.* New York: Norton.

Watson, G., ed. (1942). *Civilian Morale.* Boston: Houghton Mifflin.

Weber, M. (1948). Religious rejections of the world and their directions (originally published 1915). In H. H. Gerth and C. Wright Mills, eds., *From Max Weber* (pp. 23–59). London: Routledge and Kegan Paul.

(1976). *The Protestant Ethic and the Spirit of Capitalism,* tr. T. Parsons. 2nd ed. London: Allen and Unwin (originally published 1905).

(1978). *Economy and Society: An Outline of Interpretive Sociology,* ed. G. Roth and C. Wittich. Berkeley: University of California Press.

White, R., and Lippitt, R. (1960). *Autocracy and Democracy.* New York: Harper.

Whitehead, T. N. (1938). *The Industrial Worker.* Oxford: Oxford University Press.

Woodward, W., and Ash, M., eds. (1982). *The Problematic Science: Psychology in Nineteenth Century Thought.* New York: Praeger.

Wortis, J. (1950). *Soviet Psychiatry.* Baltimore: Williams and Wilkins.

Yates, F. (1966). *The Art of Memory.* London: Routledge and Kegan Paul.

Young, R. M. (1966). Scholarship and the history of the behavioural sciences. *History of Science,* 5, 1–51.

Index